Today There Is No Misery

Today There Is No Misery

The Ethnography of Farming in Northwest Portugal

Jeffery W. Bentley

The University of Arizona Press   Tucson & London

The University of Arizona Press

Copyright © 1992
The Arizona Board of Regents
All Rights Reserved

⊛ This book is printed on acid-free, archival-quality paper.
Manufactured in the United States of America

97 96 95 94 93 92   6 5 4 3 2 1

Library of Congress Cataloging-in-Publication Data

Bentley, Jeffery W. (Jeffery Westwood), 1955–
    Today there is no misery : the ethnography of farming in northwest
Portugal / Jeffery W. Bentley.
       p.   cm. — (Arizona studies in human ecology)
    Includes bibliographical references and index.
    ISBN 0-8165-1244-2 (cloth)
       1. Agriculture—Portugal—Pedralva (Braga) 2. Agricultural
innovations—Portugal—Pedralva (Braga) 3. Ethnology—Portugal—
Pedralva (Braga) 4. Farmers—Portugal—Pedralva (Braga)—Economic
conditions. 5. Farmers—Portugal—Pedralva (Braga)—Social
conditions. 6. Agricultural ecology—Portugal—Pedralva (Braga)
7. Pedralva (Braga, Portugal) I. Title. II. Series.
S469.P82P433 1992
306.3′49′0946912—dc20   91-24390
                                                              CIP

British Library Cataloguing-in-Publication Data
A catalogue record for this book is available from the British Library.

Dedicated to my parents, Craig B. Bentley and Verlyn Westwood, to my grandfather Le Roi Bentley, and to the memory of Vere Westwood and Bessie Burlington Bentley

Contents

Illustrations

Tables

Acknowledgments

The people of Pedralva often characterized the period before 1964 as one of extreme, miserable poverty. Emigration and changes in agricultural technology described here are partially responsible for eradicating agonizing want, and the villagers often told me that "today there is no misery" (*hoje não há miséria*).

My study was of the farm life of one parish in the Minho funded by the USAID PROCALFER project, based in Lisbon, through the Department of Agricultural Economics, University of Arizona, and the Policy and Economics Study Team (team members included Jimmye S. Hillman, Roger Fox, Timothy Finan, Scott R. Pearson, Timothy Josling, Mark Langworthy, Francisco Avillez, and Eric Monke). The Tinker Foundation provided a supporting grant. Special support was given through Manuel José de Almeida, James Black, and Phil Warnken of PROCALFER.

An earlier version of Chapter 4, titled "Kinship, Inheritance and Land Fragmentation in the Minho (Portugal)," was presented at the thirteenth European Congress of Rural Sociology, Braga, Portugal, April 1–4, 1986. Travel to that meeting was partially funded by a Graduate Student Development Fund Grant, Graduate College, University of Arizona. Another version of the same chapter was published in *Human Ecology*. Chapter 5 is revised from an article published in the *Journal of Forest History*.

Euriço Sérgio Oliveira, of the Repartição das Finanças, Braga, shared the results of a newly completed land survey with me, from which much of the land data in this study are derived. Amadeu Rodrigues Ribeiro and Carlos Manuel Soares Morais, of the Cooperativa

Agrícola e Leiteira de Braga were very supportive of this study and allowed me to copy milk production data from their archives.

Raúl Iturra, Brian O'Neill, Mary Bouquet, Alice Geraldes, and Alberto Alarcão shared their hospitality and insights with me in Portugal. Catherine Besteman helped in the field work and spent many hours discussing it with me, in Portugal and Arizona.

Walt Allen helped with the computer equipment at the Department of Anthropology, University of Arizona. Glenn D. Stone gave hours assisting with the statistical analysis, data management, aerial photographs, and the university's maniframe computers. Jonathan Mabry helped prepare the map in Chapter 5, which Samuel Trigueros later redrew. Ana Isabel Acosta drew the figures in Chapter 4. Darlan Matute drew the maps in Chapter 1 and redrew some of the other figures. Norm Shrewsbury converted my color slide photographs to black-and-white prints. Elizabeth Bentley helped checked the numbers in the tables.

Robert Netting, Brian Juan O'Neill, Susan U. Philips, Roger Fox, and Craig Bentley read an earlier draft of this book in manuscript. Mary Bouquet, David J. Siddle, and Theodore Downing read an earlier draft of Chapter 4, and Alice Ingersoll read an earlier draft of Chapter 5. All made useful comments.

Keith L. Andrews and the Crop Protection Department of the Escuela Agrícola Panamericana, El Zamorano, Honduras, provided office space and computer facilities for editing the manuscript.

To all of these people goes my most heartfelt thanks, with special thanks to Elizabeth Bentley for her encouragement while I was writing and editing this book.

It has been my pleasure to have worked with Robert M. Netting over the course of this study, benefiting from his scholarly knowledge, encouragement, and good common sense. We had many long talks in Portugal and after I returned from the field.

I am most grateful to the people of Pedralva, who endured my census, interviews, and many questions. I owe them my thanks for treating the intruder as a guest and for allowing me to see something of their lives. They are essentially private people and have enriched my life through the example of their dignity, resourcefulness, and goodness.

Today There Is No Misery

1. A Parish in the Minho

Pedralva, a parish (*freguesia*) in northwest Portugal has felt profound changes since the mid-1960s and has generally adapted to them gracefully. By Redfield's (1960:18–19) definition of peasants as rural agriculturalists who do not farm for profit, but who are linked to market towns and urban centers as a class segment of a larger society, the parish was a peasant community until the mid-1960s. Almost everyone worked in agriculture, although over half had little or no land. Parishioners lived on the maize, beans, and wine they produced. After 1964 so many men emigrated to France or took other off-farm jobs that Pedralva became a suburbanized[1] community of worker-peasants (Holmes 1983) who were only partly dependent on agriculture for their livelihood. Pedralvans became more firmly attached to Portuguese national society through off-farm work and emigrant wages saved in Portuguese banks, and by selling milk and other agricultural products. The agricultural and nonagricultural sectors have grown so rapidly that while agriculture made up a smaller percentage of the total local economy in the 1980s than in the 1960s, agricultural production was much higher.

In much of this book I deal with intracommunity behavioral differences related to unequal access to land and cattle in 1984. Community members simultaneously hold notions of social egalitarianism and social rank, both of which are expressed in ritual and symbolic contexts. Local notions of social stratification are reflected in the tension between rural groups and are consistent with the different behavior of different households; for instance, large farmers often have tractors, but small farmers rarely do. Large farmers have larger houses, eat more costly food, are more likely to have cars, and are almost the only ones who educate their children beyond primary school. As will be seen, their agricultural technology is less labor-intensive.

Davis (1977) laments that of forty-five to fifty studies of stratification in the Mediterranean, only five give enough economic detail to measure stratification accurately in quantitative, economic terms. One of the best studies is Cutileiro's (1971) description of socioeconomic stratification in southern Portugal. Cutileiro demonstrates that Vila Velha, in Portugal's Alentejo Province, is characterized by gross wealth and status differences. Land for growing wheat is owned by relatively few large landowners, with some fields owned by many small farmers, but most of the people are landless laborers. Cutileiro paints a picture of rural misery, as the landless are totally dependent on the wealthy for underpaid seasonal jobs. There is almost no personal social contact between members of the different classes. Wealthy women may chat with their maids to learn village gossip, and upper-class men may exploit their workers' wives for sexual favors, but there is little other social contact between the two groups.

Although northern Spanish and Portuguese communities are generally less stratified than the one Cutileiro describes in southern Portugal, some ethnographers gloss over economic differences within the community, as when Christian (1972:19) states that in northern Spain "a man with seventy cattle lives, to all outward appearances, in precisely the same fashion as the man who has eight." Some ethnographers of northern Spain have described inequality, but in largely descriptive terms. Douglass (1969) explains how impartible inheritance among the Spanish Basques insures the maintenance of viable farms but disenfranchises most offspring. Iturra (1980)[2] describes the systematic disinheritance of most of the siblings in a Galician community.

In a carefully documented study of a small village in northeast Portugal's Trás-os-Montes, O'Neill (1984) demonstrates vast socioeconomic differences within the community, linking wealth and status differences to landownership, social groups, cooperative labor, kinship, and inheritance. Villagers recognize three social groups: (1) *propietários*, (2) *lavradores*, and (3) *jornaleiras* and smallholders. Propietários (large owners) with thirty to fifty hectares of land form an elite group of four households. There are twenty-one households of lavradores (farmers) with six to thirty hectares of land, and thirty-one households of smallholders and jornaleiras (day laborers, in this case all females) with six hectares or less. Most jornaleiros own only small garden plots.

Since wealth and status differences are much greater in southern Iberia than in the North, the North is comparatively egalitarian. Nonetheless, northerners have created a polite fiction of rural egalitarianism in a social environment of real economic differences. For example, a wealthy peasant household in Pedralva owned two farms but rented one to sharecroppers and retained the right to harvest some of the wine grapes on the rented farm. The rented farm had two houses, one where the sharecroppers lived and an empty one that the owners had recently remodeled for the future use of their children. In October 1983 the owner traveled to the rented farm for the grape harvest. He brought most of his household members and a number of other people who owed him favors. Although the farmer had already assembled about seventeen workers, the sharecropper spent the day helping with the harvest. The farmer's daughter prepared and served lunch in the remodeled house. The farm household members and workers washed up in a common basin, and all ate at the same table. There were not enough wine bowls for everyone, so every four people were given a vessel to share, big white china bowls with bright floral and abstract designs on the outside. Both wearing patched work clothes, the farmer and the sharecropper sat side by side and drank from the same bowl. After serving himself, the farmer would turn to the sharecropper and politely say, "Would you like some of this, Senhor José?" By working, washing, eating, and drinking together, and by addressing each other respectfully, the wealthier Minhoto peasants shorten the social distance between themselves and poorer neighbors.

Much of social life in northern Iberia follows this form: wealth and status differences are muted by cultural symbols of personal interaction, expressed publicly in Goffmanesque (1959, 1967) rituals of deference and demeanor. Symbols of inequality are often expressed covertly, by gossip, innuendo, and backbiting. The northern rural folk are at once egalitarian and socially stratified, perfectly aware of socioeconomic differences that affect individual status but holding contradictory ideals of cultural and moral equality and socioeconomic status differences. The statement heard so often in the North that "aqui somos todos iguais" (here we are all equal) refers to cultural equality, not socioeconomic parity. Cultural equality means that all adults are worthy of respect and are free to enter into and leave contracts. I once observed a farmer ask two women if they wanted to come work for her one day. Instead of asking directly, she said, "We will be planting potatoes on Monday." The workers replied that they were too busy. The farmer persisted, mentioning other days when she needed workers. One worker finally ended the discussion by saying, "Eu também tenho a minha vida" (I have my own life too).

The patterns of agrarian ecology, agricultural technology, land fragmentation, and land use changes are different for each household in the parish and can be largely related to socioeconomic differences, especially farm size. Previous descriptions of farming in northwest Portugal have tacitly assumed no important differences in farming behavior in Minhoto communities (Lourenço and Alves 1968; Oliveira, Galhano, and Pereira 1983).

In the remainder of this chapter I will outline the natural and social environment of Pedralva, a parish in the *concelho* (borough)[3] of Braga, *distrito* (district) of Braga, in the center of the Entre-Douro-e-Minho (the Minho) Province of northwestern Portugal (see Figures 1.1 and 1.2).

Pedralva comprises about ten square kilometers[4] of hilly woods and cropland on the eastern edge of a plateau above the Cávado River valley. In Table 1.1 the distribution of Pedralva's forest and fields is shown. The parish is nearly three-fourths wooded, and farmers own most of the land. Most of the parish lies between three hundred and four hundred meters above sea level. Pedralva is about fifteen kilometers from Braga, the historic capital of the Minho and a city of 70,000 inhabitants. In 1984 Pedralva had 1,109 residents, in 261 households.

The parish's major products are fresh milk, maize, wine, rye, grass, and potatoes. Minor crops of olive and fruit, beans and vegetables are also grown. In addition to Holstein dairy cows (*vacas turinas*), a local variety of work cow (*barrosão*), horses, sheep, goats, chickens, pigs, and rabbits are also raised. The dominant forest plants are maritime pine (*pinheiro bravo* [*Pinus pinaster*]), eucalyptus, oak, chestnut, gorse, ivy, blackberry, broom, heather, and ferns. The trees are used for lumber and firewood. The brush plants (gorse, broom, heather, ferns, and others) serve as stall bedding for animals. About two thousand millimeters of rain falls per year, mainly in the winter, with little annual variation (Stanislawski 1959:39). Irrigation water for summer comes from springs, wells, and a stream.

Two bus lines make the half-hourly trip to Braga several times

FIGURE I.I. The Provinces of Portugal

daily, and many people, especially young men, ride motorcycles to work. Most rural Minhoto parishes are close to cities. The entire suburbanized region is densely populated, with over 250 people per square kilometer in the central Minho (Guichard 1982). Agriculture now makes up an important minority fraction of the central Minho's total economy. Farming represents economic security: land provides a "safety net" for people without secure job tenure.

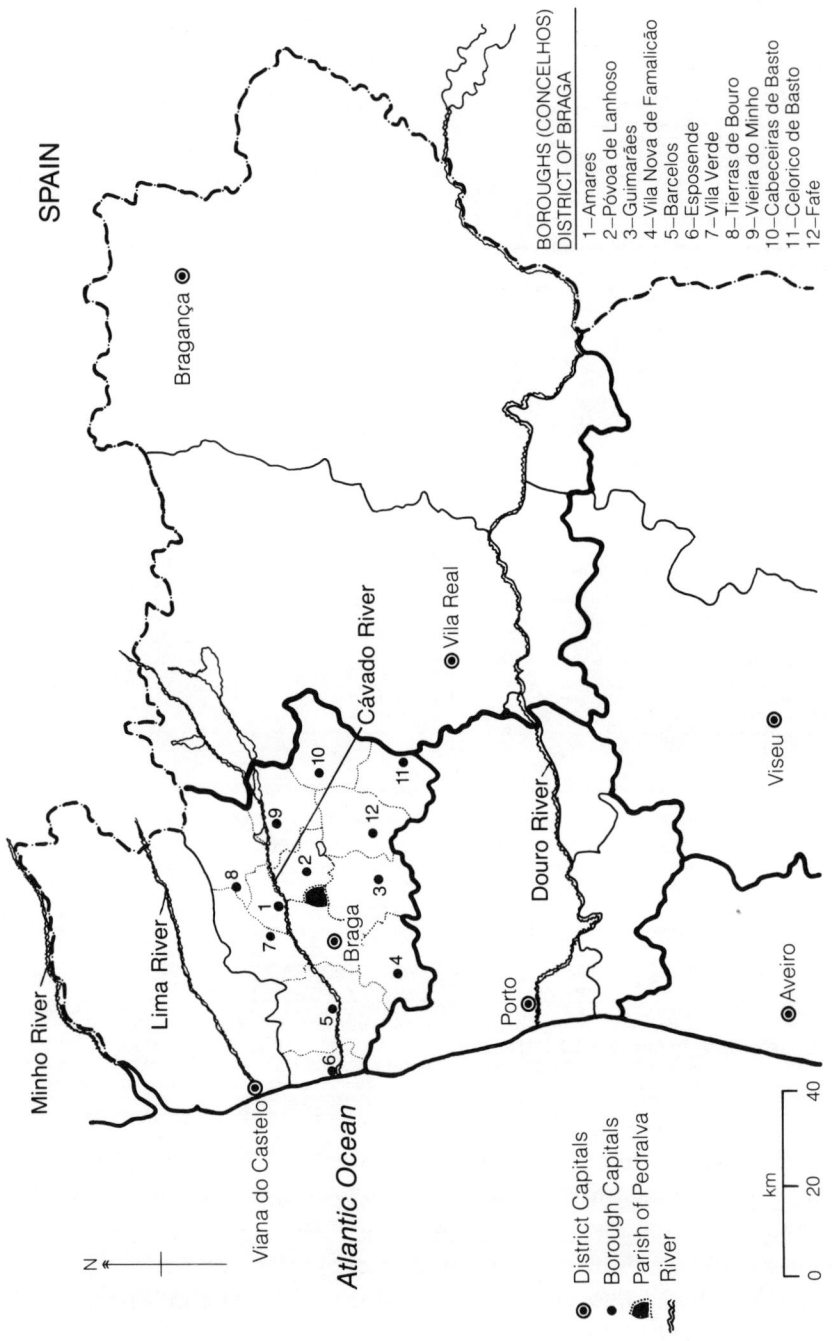

FIGURE 1.2. District of Braga

BOROUGHS (CONCELHOS)
DISTRICT OF BRAGA

1—Amares
2—Póvoa de Lanhoso
3—Guimarães
4—Vila Nova de Famalicão
5—Barcelos
6—Esposende
7—Vila Verde
8—Tierras de Bouro
9—Vieira do Minho
10—Cabeceiras de Basto
11—Celorico de Basto
12—Fafe

◉ District Capitals
● Borough Capitals
◆ Parish of Pedralva
〜 River

SPAIN

Minho River
Lima River
Cávado River
Douro River
Atlantic Ocean

● Bragança
◉ Vila Real
◉ Viseu
● Viana do Castelo
◉ Braga
Porto
◉ Aveiro

N

km
0   20   40

Table 1.1. Distribution of Fields and Forests (All land figures are in hectares)

Owners and Renters[a]	Number	Fields	Forests[b]	Total Fields and Forests
Resident landless	82	0	0	0
Resident nonfarmers	98	16.397	51.3140	67.7110
Farmers[c]	79	148.078	352.8590	500.9370
Parish common lands		0	9.3400	9.3400
Emigrants to Europe[d]	40	2.935	17.8205	20.7555
Emigrants to Brazil[e]	6	0	3.7800	3.7800
Sobreposta[f]	21	2.755	24.9800	27.7350
Others[g]	46	6.603	90.6090	97.2120
Totals	372	176.768	550.7025	727.4705

[a]Land that is rented is listed with the household that rents it in, rather than the one that owns it and rents it out.

[b]Includes a few moors, meadows, and rocky outcrops.

[c]Households that keep at least one cow, including eighty-one resident farm households and a farm owned by a physician in Braga (born in Pedralva) who visits nearly daily. Three farms are excluded because all their land is in other parishes; they are in the remote village of Carvalho, divided among Pedralva and three other parishes.

[d]Most of these emigrants are in France. Of their 2.935 hectares of cropland, 0.288 are in fallow. During a brief visit in 1986 I learned that 0.118 hectares were being worked by the emigrants themselves, who had just returned. The remaining 2.529 hectares are unaccounted for and are probably loaned to relatives. The emigrants own an additional 6.518 hectares of cropland not listed here, of which 2.474 hectares is rented to farmers and 4.044 hectares to nonfarmers. More land in this category is rented (or loaned) to nonfarmers than to farmers because emigrants generally come from the former *pobres*, and loan or rent plots to kin.

[e]Two emigrants jointly own 0.26 hectares of cropland, rented to a nonfarmer.

[f]The parish bordering Pedralva on the west. Six residents own cropland in Pedralva.

[g]Residents of other nearby parishes. Over two hectares is fallow and just over four is a large, recent clearing, most of which is not yet farmed, discussed in Chapter 5. This group owns 13.778 hectares of fields, mostly rented out, including over 0.560 hectares to a large farmer and 5.336 hectares to sharecroppers.

Pedralva's unequal land distribution, as in Trás-os-Montes, is crucial to social group ascription. The number of social groups, and their local names (emic categories), varies from community to community in northern Portugal. Guerreiro (1981) mentions lavradores (farmers) and *cabaneiros* (cottagers) in the Serra do Barroso, in Trás-os-Montes near the upper Minho. Pedralvans state that in a neighboring parish there were two social groups before the 1960s: propietários (landowners) and cabaneiros. The three propietários owned all the land, the cabaneiros owned none. After the 1960s the cabaneiros used money earned as emigrant workers to buy out the landowners, eliminating the highest-ranked group.

In the early twentieth century there were three social groups in Pedralva: propietários, lavradores, and *pobres* (landowners, farmers, and the poor). There were, and still are, no nobles or bourgeois. Propietários owned several farms, delegating the physical labor to share-croppers or hired hands. The farmers owned less land, some of which might be worked by sharecroppers, but the farmers generally tilled at least some of their own soil. Most people were landless, or nearly so, and worked as sharecroppers (*caseiros*),[5] as agricultural day labor-ers (jornaleiros), and as servants (*criados*) for the farmers and large landowners.

Older people say that the propietários sold their large estates by the 1920s, leaving only lavradores and the pobres. In the mid-1960s France opened its doors to emigrant workers, and men from Pedralva poured across the Spanish border hoping to reach France. In the early years emigration was largely illegal and therefore clandestine. Portu-guese and Spanish authorities arrested men attempting to emigrate, but once in France they could "walk into the arms of the gendarme without danger."

Many of the workers who went to France were pobres, sharecrop-pers, and day laborers. The flood of emigrants beginning in 1964 changed traditional patterns of labor use and social relations dramat-ically. In the 1960s there were about twenty sharecroppers in Pe-dralva. By 1984 there were only six, similar to the pattern that Caldas (1981) observed for another Minhoto community.

Villagers speak of two times: *antes da emigração* (before emigra-tion, i.e., before the mid-1960s) and *depois da emigração* (after emi-gration).[6] By all accounts, emigration brought great change to the region. After the 1960s emigrants emerged as a new social group (Brettell 1979, 1983), and many are now among the wealthier villag-ers. Some emigrants, however, never visit Portugal. They are bitter about the poverty of their youth and wish to stay in France, saying, "Portugal ate my flesh, but will not eat my bones." Other emigrants live abroad eleven months of the year with their entire households and return to Pedralva each August to a large modern house that stands empty most of the year. Many hope to retire in Pedralva, al-though their children are not eager to leave Paris or Marseilles.

Still others are married men who go alone to France. Their wives stay in Pedralva to tend their small farms or fields, seeing their husbands once a year. One elderly woman recalls not seeing her husband at all for seventeen years because he used his August vacation to take a second job in France. The separation was particularly painful because both are illiterate and could not even write to each other.

After emigration the relationship between the social groups changed. While there were now three social groups (farmers, emigrants, and the poor), the social boundaries blurred. The poor largely left agricultural work. Those who found local off-farm work in the expanding economy were still poor, but not like before. Some emigrants stayed abroad for many years, carefully saving their wages, while others stayed for a few years and did not save enough money to build large houses. Some saved enough money abroad to buy land and became farmers. Others were already farmers when they left and invested their savings in agriculture.

No one agrees on exactly what constitutes membership in any of the groups. Before emigration a person with two cows (a plow team) was a farmer, or a plower, the literal meaning of lavrador. After 1978 most of the work cows were replaced with dairy cows. Now most plowing is done by tractors, and cows are kept for milk. A household with even one cow can join the local dairy cooperative and sell fresh milk. Some people with only one cow claim to be farmers, albeit poor farmers. Others claim that a single cow does not make one a farmer. The parish priest declared that there were only thirty farms in the parish, but that "there are many who say they are farmers, but they are not."

Because of the hardship of saving money on a poor person's earnings few Pedralvans have ever moved up the socioeconomic ladder without emigrating; getting enough cattle through sheer muscle and sweat to become legends in their own kine. The most celebrated case in Pedralva was a man I will call O Pote.[7] O Pote was born in 1919 and married in 1944. His wife and he were both from small farm families. Between the two of them they inherited only half a hectare of land, a house, and a little forest parcel, but by the time I knew him in 1983–1984 he had nearly four hectares of fields and over two hectares of forest, in twenty different parcels, as well as nine dairy cows and a thriving business making and selling bagaceira, the local brandy distilled from grape skins.

Minhoto peasants are tremendously hard workers, O Pote was out-standing even by local standards. One villager explained that O Pote started his socioeconomic climb with a nanny goat. The goat got pregnant and had two kids. O Pote sold the three goats at a fair and bought a calf, which he raised. When the cow produced another calf he raised it and trained them both to work as a team. He did custom plowing and other work with the cows, saving the money, and also sold firewood. They say he would load his cart with the fuel in the daytime, then get the team to pull it the fifteen kilometers to the regional capital, Braga, while he slept on the load, so that he could work even as he slept.

O Pote's success was based in part on his skills of persuasion. In one instance he argued in favor of land consolidation to swap a little garden for a much larger field. His little house garden was next to his next-door neighbor's, so O Pote had a right-of-way through the neigh-bor's vegetable patch. The neighbor also had a field two kilometers away that was three or four times the size of O Pote's garden. O Pote convinced him to swap, pointing out that the neighbor would avoid a long trip to the field (a trip O Pote was more than willing to take), would enlarge his garden with the addition of O Pote's, and would now be able to plow up the right-of-way.

O Pote invested all his money in production: land, distilleries, and a sawmill (which he later sold). It was not until 1981 that he used some of his savings to repair the old house.

Villagers commonly distinguish between large farmers (*grandes lavradores*) and poor farmers (*pobres lavradores*), although there is little consensus as to what distinguishes the two groups. When asked what a large farm is, people usually define the class by example, pointing to one of the three largest farms in Pedralva. As Rosch (1978:35–40) observes, it is often easier to define the prototypical member of a category than the category's boundaries. In Table 1.2 the unequal distribution of cropland (including fields and gardens but not forest) and cattle in Pedralva is shown by household. I have arbitrarily divided landholdings into six categories: no land; less than half a hectare; one-half to one hectare; one to two hectares; two to four hectares; and more than four hectares.[8] A holding includes land that is owned or rented in. These etic categories have the advan-tage of clear boundaries, while the emic ones have fuzzy edges, but they do correspond roughly to the emic categories of social groups. The pobres are in the first two categories; the pobres lavradores in

Table 1.2. Distribution of Cropland and Cattle by Household

Farm Class (in ha)	Number of Households	Average Herd Size	Average Size of Holding (in ha)
No land	95	0	0
0.0001–0.4999	84	0.060	0.164
0.5–0.9999	28	1.321	0.750
1–1.9999	22	3.091	1.406
2–3.9999	16	5.750	3.043
4 and above	8	13.375	5.712
Total[b]	253	1.221	0.619

[a]As mentioned earlier, one farm in Pedralva is owned and managed by a physician who lives in Braga but was born in the parish and inherited the farm. This farm is included in this table, in Chapter 4, and in other discussions of agriculture. The farm has no resident household members and so is excluded from the remaining tables in this chapter and from other sociological discussions of the parish's 261 households.

[b]Nine households deleted from sample because of insufficient data.

the third category. The fourth category is made up of definite lavradores, while the grandes lavradores are all in the fifth. Emigrants are included in all categories.

Cattle distribution corresponds closely to land distribution (see Chapter 4), and both are extremely unequal. Of 262 holdings, 179 (68%) are below half a hectare. The upper 10 percent of the households have 60 percent of the farmland (see Figure 1.3). While all farms in Pedralva are small, some are much smaller than others.

The sociological implications of uneven land distribution are illustrated in Tables 1.3–1.5, with herd size (instead of amount of land) to indicate farm size. Using cattle instead of land has several advantages. Cattle represent the relationship of the household to the market economy, since milk is the main cash crop. Using the number of cattle as an independent variable controls for differences in land quality, since more cattle can be supported on better soil. Also, cattle data are more complete than land data.[9]

A few households have enough land to support all members as full-time farmers, but most do not. Most households make their living by a mixed economic strategy, with some members working off-farm, while others work at home on the farm or garden. The less land a household has, the more of its labor must be sent off-farm, either locally or abroad.

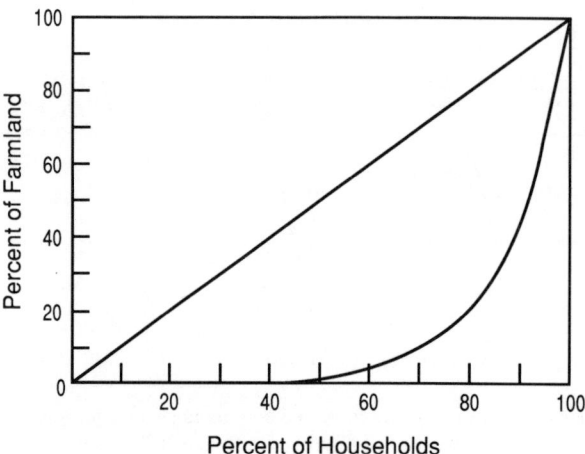

FIGURE I.3. Lorenz Curve of Parish Farmland

Off-farm wages are much lower for women than for men (see Chapter 3). Women are also denied access to the most highly paid off-farm work, in the construction trade. A household with some land, and male and female members, maximizes the value of household labor by sending males to work off-farm, while the women work on-farm. Household members pool their salaries and farm produce, sharing more or less equally in the consumption of household resources. Those with emigrant male members usually spend their savings on items that all members can enjoy, such as a new house or indoor plumbing.

Men and women—especially husbands and wives—form close, long-lasting economic bonds. Generally neither a man nor a woman could survive without the other. Men employed in local construction work earn the equivalent of just over U.S. $100 (1984) a month, not a comfortable living wage. Men depend on the garden produce that their female kin produce, and on their domestic services (e.g., cooking, washing) to supplement their wages. Women depend on their male relatives for cash to buy consumer goods and agricultural supplies such as seeds and fertilizer, and to help with the farming on Saturdays, holidays, and in the evenings. Villagers express interdependence of both sexes by comparing each to the wheel of a cart: "uma roda só não anda" (one wheel by itself does not roll). This sexual division of labor helps villagers adapt to the economic envi-

Table 1.3. Off-Farm and On-Farm Workers, Male and Female

A. Average Number of Workers per Household, by Sex and Location

Herd Size	Number of Households	Females Off-Farm	Males Off-Farm	Females On-Farm	Males On-Farm
0	189	0.249	0.730	0.370	0.101
1–2	33	0.182	1.030	1.227	0.455
3–4	18	0.167	0.667	1.722	0.778
5–7	14	0	0.571	1.786	1.571
8–33	7	0.143	0	2.143	2.286
Total	261	0.218	0.736	0.655	0.330

B. Absolute Number of Workers by Farm Size, Sex, and Location

Herd Size	Number of Households	Females Off-Farm	Males Off-Farm	Females On-Farm	Males On-Farm
0	189	47	138	58	19
1–2	33	6	34	42	15
3–4	18	3	12	31	14
5–7	14	0	8	25	22
8–33	7	1	0	15	16
Total	261	57	192	171	86

ronment of small rural agricultural holdings and discrimination against women in the urban labor market.

The sexual division of labor is different for each farm size (see Table 1.3). The few females involved in off-farm work come, almost entirely, from households without cattle. These are the poorest, smallest holdings, with the lowest on-farm labor requirements. Women usually work off-farm only when they have very little land of their own to work. Almost four times as many men have local off-farm jobs as women. Like the women, the men are more likely to work off-farm when they have little land. On the average, the smallest size of holding has fewer males working off-farm than does the next largest size, because the smallest holdings have the fewest household members.

Three times as many women as men work on household farms in Pedralva. A woman commonly gets up at 5:00 in the morning to pack a hot lunch for her husband to take to his construction job in Braga, then makes his breakfast. After sending the children to school, she works in her fields, interrupting her farming day to wash clothes, clean the house, and cook meals, then finally do the supper dishes while the men watch television or step out for a drink. She may take her smallest children with her to the fields or leave them at home with an elderly household member. In larger households, with several adult women, those over the age of sixty generally prefer to stay home and handle domestic tasks while younger women take on the farming chores. Women tend to work longer days than men, often taking little or no rest time at all, except for a nap on Sundays.

Larger farms have more women and men working full time in agriculture. However, while the average number of female workers per farm increases slightly with farm size, the average number of male workers rises steeply with farm size. Households with at least one cow keep female members on the farm, where their labor earns higher returns than they would earn selling their labor in town. A household must have about five cows before the opportunity cost of sending males off-farm equals the value of their work on-farm. Many of the men who work the smallest holdings are retired emigrants, essentially hobby farmers with few other opportunities for their labor.

Households with no cattle employ an average of 0.307 women on-farm, and 0.249 women off-farm, for a total of only 0.556. Many of the women in this category listed their occupation on the census as "homemaker" (*doméstica*) and so are not coded as off-farm workers or on-farm workers. Many of these women spend much of their time cultivating a small garden and tending small animals, essentially agricultural activities but on such a small scale that they do not consider themselves to be farmers.

Full-time male farmers generally work on the largest farms, because men with little land can make more money in town than in farming. Women work off-farm only when they have nearly no land at all, because their labor is not as well paid off-farm and because they are burdened with young children and elderly dependents. While men will do any agricultural or gardening task, and will work with the animals, they will not do housework. This rigidly observed cultural ideal creates an added demand for female labor at home.

In Table 1.4 the emigration experience, both for different-sized farms and for males and females, is outlined. Women were absent as

Table 1.4. Average Number of Years of Emigration per Household

Herd Size	Number of Households	Average Years of Female Emigration	Average Years of Male Emigration	Percentage of Emigrant Households[a]
0	189	1.593	5.952	0.13
1–2	33	0.667	9.182	0.15
3–4	18	0.778	13.167	0.33
5–7	14	0	7.143	0.07
8–33	7	0.286	6.000	0
Total	261	1.299	6.923	

[a]Percentage of households in each category with at least one member currently out of Portugal.

emigrants for an average of just over one year, while men were gone for a mean of nearly seven years. Because women are locked out of highly paid blue-collar jobs, men often emigrate without their wives. Many of Pedralva's sons have worked abroad, generally as construction workers in France. Women who emigrate tend to join husbands already abroad.

There is not much difference in average number of years of male emigration per household by farm size. Most categories have had men absent for six to ten years. The exceptionally large figure of 13.167 years for the three to four–cow farm category reflects the extremely long emigration experience of one man, who was absent for fifty-seven years (1922–1979).[10] When his time is subtracted, the average number of years of male emigration for this category falls to ten.

There is only a slight tendency for middle-sized farmers to have more emigration experience than either large farmers or households with little or no land. Larger farmers do emigrate sometimes: several of the current large farmers emigrated for several years in the mid-1960s, dodging military service in the colonial wars of Angola, Guinea-Bissau, and Mozambique; other large farmers emigrated to save money for agricultural investment. Landless men or those with very little land emigrated and used their savings to buy land, acquiring the status of small farmers. This is reflected in the slightly higher emigration experience of farmers with one to two and three to four cows.

Table 1.5. Average Household Size by Farm Size

Herd Size	Number of Households	Mean Size
0	189	3.831
1–2	33	4.788
3–4	18	5.500
5–7	14	6.214
8–33	7	5.857
Total	261	4.249

Emigration experience is lowest for households without cattle, because that category includes a higher proportion of solitary or elderly people, retired couples, and young couples who have not had the opportunity to emigrate. Men born after 1955 have not been able to emigrate because of the world recession after 1973. Also, many of the people with no land at all emigrated permanently, or else the husband and wife emigrated together. Female emigration is highest in the nonfarm category because a landless woman has the least reason to stay in Portugal and is freer to join her husband in France. Like women who work off-farm locally, women who work abroad generally work in the service sector, often as cleaning women. Those who accompany their husbands to France supplement the household income there. But since the men who go abroad live in barracks or in apartments with other men, couples must find more-expensive housing.

Emigration from Pedralva has been so high it has transformed the social landscape, bringing more money into the community, allowing landless laborers to leave farm work, and encouraging farmers to buy labor-saving machinery.[11] Farmers once depended on sharecroppers, household servants, and day laborers to provide labor at extremely low wages. Their relationships often lasted for a lifetime and were important social and economic ties. There are now far fewer sharecroppers, household servants, and day laborers than there were before emigration. In the mid-1960s many people took the opportunity to escape these oppressive social and economic relationships by emigrating.

Landless households (see Table 1.5) in Pedralva include solitaries, elderly couples, and very young couples. The larger farm households include more three-generation households, more resident celibate siblings, and some household servants—generally elderly servants who have been with the family since childhood. Generally, the larger holdings have the greatest number of household members. This reflects different postmarital residence patterns for large and small-holdings. Large farmers generally form stem family households, in which one son or daughter (the major heir) marries and brings his or her spouse to live in the parental household; other household members include the unmarried children of both these couples (Willems 1962). In contrast, poor people usually reside neolocally (newlyweds find separate housing from their parents). These results are compatible with Netting's (1982b) cross-cultural and historical survey, which found that the greater labor requirements and greater capital resources of wealthy households led to a larger household size among the wealthy than among poorer households in the same communities.[12]

Pedralva is not an egalitarian peasant community; so much of the work force now works off-farm, both locally and abroad, that Pedralva is a worker-peasant community. The parish has important traditional socioeconomic differences, related directly to access to land. Much of the change since 1964 has been away from labor-intensive, capital-scarce farming toward greater market orientation. Earlier farming relied on enduring social relations and the almost exclusive use of local resources. As will be shown, the growing importance of migration, money, machinery, milk, market, and agrochemicals have turned Pedralva away from an ecologically sound, sustainable agriculture and have also ended a poverty so bitter that the people called it misery. Economic differences are now keyed into behavioral differences, because each farm family lives and works in a unique natural and economic environment.

2. Agrarian Ecology, Cultural Homogeneity, and Behavioral Diversity

If culture is to behavior as language is to speech (Goodenough 1971: 19), then people can share a culture but behave quite differently, without being deviant. The people of Pedralva share a common culture. The local agricultural technology is part of even nonfarmers' cultural heritage. Although composed largely of nonfarmers, the folk dance troupe from Pedralva displays the agricultural implements of a generation ago while they dance, as symbols of the farm legacy of the entire community. Although most Pedralvans are not farmers, all understand the local crops and domesticated animals. Almost everyone has worked in agriculture from time to time, if not as a paid worker then as a favor to a friend, kinsman, or neighbor. Many farm tasks (e.g., potato harvesting) are also done in the household gardens, which nearly everyone has. A nonfarmer with half a hectare of land may grow potatoes, wine grapes, grass (for rabbits), beans, fruit trees, and sometimes maize (for chicken feed or bread), besides garden vegetables. Forest products such as firewood, gorse, and stone are important for nonfarmers and farmers alike.

Even the returned emigrant, who expresses his new status and relative wealth by constructing a tile-covered, French-style suburban house, often spends his first spring back home planting potatoes, kale, and vegetables in the household garden, to have good produce on hand, to stretch his savings, to live up to the community ideal of hard work, and because it is something he knows well.

Men and women share farming knowledge. Unlike washing, cooking, and other housekeeping chores, which Pedralvans consider to be women's work, most farm tasks are seen as acceptable work for both sexes. Although women generally garden and care for animals (those jobs closest to home), men do not murmur about having to feed the rabbits or work in the garden. Women avoid climbing the tall, thin, unstable ladders to prune vines or to make haystacks (because it is dangerous), although they will climb ladders to harvest grapes.[1]

Although agrarian knowledge is part of every community member's culture, no two farms or gardens are worked exactly alike. The "typical systems" that agricultural economists like to write about are merely abstractions. Most farm tasks (e.g., planting maize, harvesting potatoes, making wine, weeding) can be done in two or more ways. Combined with an ever changing technology, there is infinite[2] but subtle variation within the farming styles of a single parish.

Local variation in agricultural technology is neither random nor idiosyncratic but is conditioned by household supplies of land, labor, and capital, which vary widely within the community. While there is an attempt to adjust household labor to the supply of land (e.g., by hiring labor or by taking off-farm jobs), the households with less land tend to have less capital and more labor per unit of land. The more land a household has, the more labor-saving machinery and agrochemicals it uses. For example, although all farmers must control weeds in maize fields, small farmers chop them with a hoe, while the larger farmers tend to use preemergent herbicides. Any farmer may use a horse-drawn cultivator to cut weeds, if he or she has access to a horse. Mere knowledge is not an important factor in intracommunity agricultural variation (Cancian 1972:125–27). Everyone knows how to use a hoe and an animal-drawn cultivator. Herbicides are newer, and less well known, but unfamiliarity is no real barrier to their adoption. Everyone knows that herbicides kill weeds, where to buy them, and who can explain their use. Small farmers—with relatively abundant household labor—do not buy herbicides because they can afford the time for hand-weeding but not the cash. Thus, agricultural techniques are part of communitywide cultural knowledge, while agricultural behavior varies for socioeconomic reasons. Culture provides a shared grounding for a range of behavioral differences, alternative agricultural technologies whose choice is determined largely by different supplies of land, capital, and labor.

Regardless of which methods a farmer chooses, all contemporary agricultural techniques are appropriate for the farms on which they are used. Few farmers adopt a new technology uncritically. Each farm household modifies the new technology, carefully adapting it to fit the farm's unique natural and economic environment (see Horton 1986). This in some ways contradicts the popular view that Portuguese agriculture is backward. Many Portuguese agronomists told me, "We have two problems, our farms are too small and our farmers are too dumb" (*as nossas explorações são muito pequenas e os nossos lavradores são muito burros*). A cover story from *The Economist* (1984:20), based on an interview with the Portuguese minister of agriculture, claimed that "the country's agricultural inefficiency is rooted in the past. Its northern estates were divided into smallholdings, many no bigger than an acre or two, as a result of the Catholic custom of sharing property equally among children. In the Alentejo region in the south, by contrast, a few absentee landlords owned large estates, some of 2,500 acres. Both regions were farmed badly." Contrary to the prevailing prejudice, Minhotos are industrious, thoughtful farmers. The techniques described in this chapter are economically rational and ecologically adaptive.

FOREST AND FIELD

The cycling of nutrients from forest to field (see Figure 2.1) takes on a pattern unique to northwest Portuguese agrarian ecology (see Stanislawski 1970:45; see Table 1.1 for the amount of forest owned by different groups). Gorse, broom, heather, fern, and other plants of the forest floor are known collectively as *mato* (brush). Brush is cut from the forest with hoes as often as every four years, although by the 1980s people cut brush less frequently (Caldas 1981). Brush may be cut at any time of the year, but more is cut in the early spring to mix with manure in the fields. Brush is stored at the farmstead, where every few days a new layer is spread on the floor of the cattle stall, forming a *cama da vaca* (lit. cow bed, i.e., stall bedding) that keeps the cows clean when they lie down and helps to soak up the urine and feces, preventing nutrient loss. The brush composts in the manure, so both are taken to the fields together when the stalls are cleaned out in early spring. Some farmers make a large mound of manure and green brush in a field of winter grass that is not allowed to go to seed and can be cleared early to plant summer maize. Layers of manure and brush are built into a mound one to two meters tall. Completely covered over in manure like an iced layer cake, the mound (*moreia*) composts the green brush, so that in a few weeks it is ready to be spread out and plowed under.

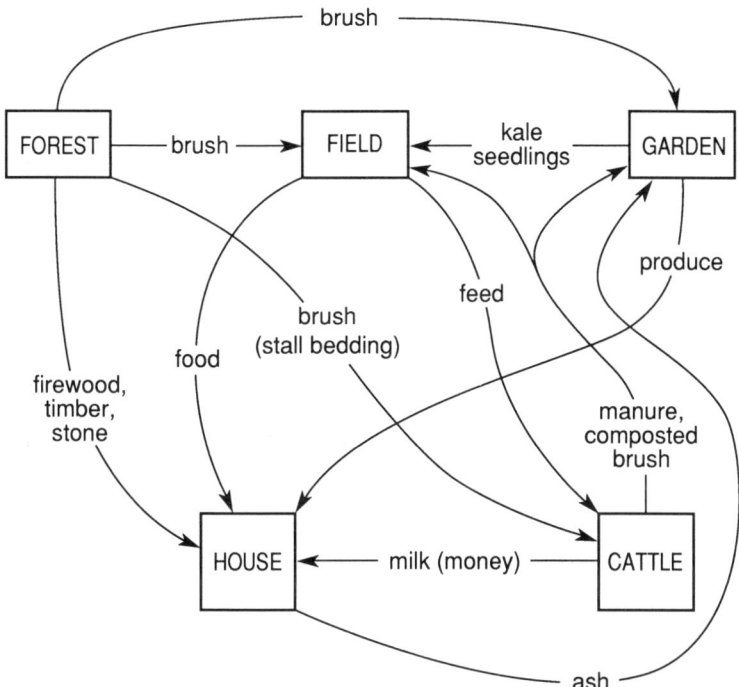

FIGURE 2.1. Material Exchanges Between Major Components of the Managed Ecosystem

Farmers say that a generation ago the brush was also used as an insecticide. When soil became infested with *bicho do milho*[3] (maize bug) they would plow as much green brush as they could into the soil, destroying the insects with the heat of rotting vegetation.[4]

Maritime pine is the most important native economic tree and the dominant forest plant, growing on all but the stoniest outcrops from the coast to the uplands. While in other parts of Iberia the pine tree is tapped for resin (Aceves 1971), in this region the trees are cut for lumber. Any forest owner can sell trees, generally through a local timber cutter who cuts and stacks the logs, then sells them to an outside timber company. Both of Pedralva's timber cutters have large farms, at least two tractors, and at least one grown son to direct and work with the hired hands. The timber cutters pay Esc (escudos) 2,400 (U.S. $18.46) per ton for pine and Esc 1,100 (U.S. $8.46) per ton for eucalyptus. Forest yields vary, but a thousand tons per hectare for pine and twenty-five hundred tons per hectare for eucalyptus are about average.

The exotic eucalyptus grows well in Portugal. It is usually hand-planted and grows to maturity in fifteen to twenty years. The eucalyptus grown in the Minho makes poor lumber; being so wet it warps and cracks as it dries. It is used for firewood or as timber for light roofs or is sold commercially for such cheap things as popsicle sticks.

A forest owner who needs lumber generally cuts a few trees himself with a chain saw. Each log can be several meters long and weigh a ton. The difficulty of finding four to six men to help cut the logs and load them onto a tractor wagon demands that the work be done at slack times. Robust young men are the most difficult laborers to recruit, because those who do not have off-farm jobs work on their own farms. A construction worker may go logging for a farmer to repay a favor or to maintain a favor-exchange relationship, but cash is seldom offered or accepted for these heavy, one-shot, irregular tasks. Home-style logging usually happens after work, which means it often gets dark before the job is done. Accidents can happen as tired men struggle in the twilight to load heavy, slippery logs onto a wagon in the confinement of a dense forest. Evening logging is an adaptation to scarce labor and new machines, although it is arguably maladaptive for the loggers. Logs are driven to a sawmill in a neighboring parish, where they are sawed into planks for a fee. A tractor and driver can be hired if the forest owner does not have a tractor.

After logging, the branches, pinecones, and treetops are usually left in the forest to dry before being gathered for firewood. Villagers claim that firewood was much more in demand before butane became popular in the 1960s. "Now no one even talks about stealing firewood," they say. Oak, chestnut, pine, and eucalyptus trees are sometimes cut and split for firewood.

Recycling farm products (manure, maize stalks, etc.) is important for maintaining soil fertility because high rainfall leaches the earth of nutrients. Cropland is also kept fertile by applying brush from forest to field, either directly or used first for stall bedding, just as Alpine Swiss farmers gather leaves, moss, and forest humus for similar purposes (Netting 1981:46–47). Wood ash from the kitchen is put on the household garden (see Stanislawski 1959:43–48; Oliveira, Galhano, and Pereira 1983:18).

While the forest is less intensively exploited than the fields, forestlands are also part of the managed ecosystem. Forest (*monte*) plots are known by three different terms, depending on their size and kind of boundary. Large forest tracts, of several hectares or more, are *coutadas*; small unwalled strips are *leiras*; and small plots enclosed by stone walls are called *bouças*. Stone fences are important, since building one takes work, marks a property boundary, and frees the forest floor of stones to let more brush grow. People also make small piles of stone (*montes de pedra*) in the forest to avoid chipping their hoe blades on rocks as they cut brush. Another labor-intensive use of forestland is planting tree seedlings, especially eucalyptus, to hasten forest growth. Until the 1960s poor people collected acorns to sell to farmers as pig feed. A few chestnuts are collected as a treat to roast at home on winter evenings, leading Portela (1981:223) to note that the Portuguese villager plays all the major evolutionary roles simultaneously: industrial worker, farmer, hunter and gatherer.

One hated native plant is the blackberry bush (*silva*), which exploits such wet, stony places as field walls and canal banks. Its vinelike branches take root, spreading as a tangled, thorny mass. Blackberry brambles are cut up with hoes and long-handled sickles, left to dry, and then burned; the ashes are plowed into the earth. A farmer who allows blackberry bushes to grow is censured in neighborhood gossip, even though the berry (*mora*) is edible. People say blackberry wine was made before 1964, although now the berry has the low status of a starvation food and people who eat it usually wait until they are not being watched.

Stone is used for many agricultural structures and implements.[5] Specialists use hand tools such as hammers, star drills, wedges, chisels, and crowbars to cut granite for building blocks and *esteios* (grape arbor posts). Cutting a straight sliver of granite twenty by twenty centimeters and five meters long, using no machines, is a truly impressive skill.

Stone was used to build almost everything before the mid-1960s. House walls are made with rectangular, megalithic granite blocks, many well over a meter long and so heavy they could be lifted with the front-end loader attachment of a tractor only if the machine was counterweighted. Lintels and doorjambs are sculpted from granite, often in elaborate scrolls. Fences, houses, churches, barns, threshing floors, and gristmills are made of rough or cut stone blocks. Village lanes and farm roads are paved with stream cobbles or blocks of granite. House floors are made of wood, but a large section of the kitchen floor is made of stone so that a fire can be built on it. The ashes are stored in a stone box (*borralheira*) behind the *lareira* (hearth), and the smoke escapes through the tile roof. Hams and sausages hang above the hearth, curing in the dense smoke. The stone hearth is the focal point of the household, and the household is the focal point of social life. A household may be referred to as a *fogo* or *lume*, both of which mean fire.

Stone as a building material has deep roots locally. Medieval sarcophagi are still visible at the ruins of two parish chapels. A three-meter-tall stone statue of a man, probably prehistoric and alone of its kind in Iberia, was found on the mountain crest above the parish and is now in a museum in Guimarães.

While the intimate relationship with stone has a long past, it does not necessarily have a long future. Cement block, brick, and multi-colored tile have become popular since the mid-1960s. Emigrants who were the impoverished children of the landless are anxious to show that they succeeded in the outside world. They display their new economic status by returning to their home community—where people remember their poverty-stricken youth—and building a large, expensive house with few native materials (see Rhoades 1979, 1980; Lucas 1983:166–68). While the farmers have begun to use cement blocks for building *cobertos*[6] (hay and grain barns), they still prefer native stone for house construction. O'Neill (1989b) also observes that farmers in Trás-os-Montes prefer stone to the emigrants' flamboyant tiles.

Parish settlement patterns are influenced by the location of fields and forests. The center of Pedralva is a field area of about thirty-five hectares, with no buildings, regarded as the best farmland in the parish. Most of Pedralva's sixteen villages surround this field area. Immediately around the villages are more small fields and gardens; beyond them lie the forests.

Houses are built near, but not directly on, the best farmland to minimize distances from farmstead to fields without taking good earth out of production. Several villages are named for nonagrarian ecological features—Souto (grove), Monte (forest), Codeçosa[7] (patch of broom), Outeiro (hill)—suggesting that they were not built on arable land. Some villages stand at the edge of the forest, where the land is steep and dry. Others are built where large boulders stick out of the ground. Some barns and threshing floors incorporate exposed bedrock in their structure. The original placement of buildings shows the careful preservation of the best land for agriculture.

FARMING SYSTEMS

The unique, complex Minhoto cropping technologies have previously been described in Portuguese, but not from an ecological perspective. Lourenço and Alves's (1968) economic study of time in agriculture provides an extensive list of agrarian routines, including the amount of time each task requires. Ethnographers Oliveira, Galhano, and Pereira (1983:17–52) describe northwestern farming systems, but much of their field work was done before 1947, to document an essentially medieval technology before it disappeared under the influence of "invading techology" (p. 7). Neither study discusses intracommunity variation.

Farmers see the tractor more as liberator than invader, freeing them from labor constraints and tedious tasks. Many of the techniques described by Oliveira and his colleagues have not disappeared in the past four decades because the preexisting farming system evolved, incorporating machinery without completely eliminating earlier techniques. Some of the older routines have been modified or have become less common and more specialized, while some new tasks have been added.

The variation in maize-growing technology is determined by each household's supply of the factors of production—land, labor, and capital—and by the ecological adaptation of each farm to a particular microenvironment. Personal or psychological factors are also important; some farmers prefer large festive work groups, while less-social farmers hire as little labor as possible. Recruiting workers has a fairly high transaction cost: people have to be visited, or word has to be sent asking them to come work. Some of these people may excuse themselves, and replacements have to be contacted. One particular farmer did not get along with any close neighbors and had to find workers from other villages in Pedralva. The farmer found it irksome to go around asking people to come work and even more annoying to do them return favors. He did not enjoy his workers' company, then complained when they left after supper, without staying to help till bedtime. But most farmers do not mind rewarding a group of kin and neighbors for a day's hard work with big meals, ample wine, and an atmosphere for good jokes and stories.

Some of the occasions that call for extra labor are maize weeding, harvest, and husking; silage cutting; potato planting and harvest; rye harvest and threshing; turning the drying hay and harvesting it;

grapevine pruning, harvest, and wine making. Most farmers recruit laborers from a small set of households, almost invariably referred to as friends. Smaller farmers often exchange labor with others (*troca de mão de obra*) to get groups large enough for some tasks. For example, representatives from six small farm households may spend a week working together one day on each farm harvesting rye. Larger farmers tend to have relationships with landless or land-poor families in which the poor either work for pay or exchange labor for *favores*, especially machinery (see Pina-Cabral 1986:158–59). A *favor* in Portuguese implies more than its English cognate; it is a good or service given to maintain a formal friendship. While there may be affective ties between friends, the exchange of economically valuable favors is the most important part of friendship.

Pedralvans hardly distinguish these types of relationships. Even paid workers eat with the farm household and may ask for favors. Exchange—labor traded for labor, machinery, or something else—is sometimes specifically negotiated (I will help you thresh rye if you will do the same for me) but more often it is not. In all cases workers are usually drawn from the households of neighbors, kin, and friends.

Labor is expressed as friendship by referring to it as help (*ajuda*) or a favor. Villagers try not to account for labor exchange too rigorously, saying, "We help each other" (*nós ajudamos os uns aos outros*) and "We don't keep track of it" (*não fazemos contas*). Ideally, friends help each other without counting the cost, loaning money, machinery, labor, and horses to each other whenever asked. Behavior comes close to the ideal; people count the cost only when they are criticized or they feel taken advantage of. One woman chided her daughter for spending too much time on the farm of the stingy farmer mentioned above. The girl retorted, closing the matter, "But didn't he come plow our field and slaughter our pig!" Another farmer complained about how he had plowed his neighbor's field, even paying for the diesel fuel himself, while his neighbor repaid him poorly by sending his son to work for only half a day. The farmer concluded that he would never again help that family.

In general most households conscientiously maintain good working relations with several others—their friends—with whom they expect to exchange goods and services over a long time. With more land and crops, larger farmers need more labor. A few farmers have full-time workers, but most meet their labor demands by hiring temporary hands at peak season and by cultivating friendships with a few land-poor households.

The man and woman who run a farm are called "the bosses" (pa-trão, patroa)[8] by the workers, but only while working on that farm. Children eat with the other kids they play with, and workers eat with the bosses at their table. Farmers generally call their workers *senhor* or *senhora*, plus their first name (Senhor Pedro, Senhora Maria, etc.), just as the workers address them. Although the farmer drives the tractor if he has one and needs to use it, members of the farm household generally do the same tasks as the workers. Land in Pedralva is demonstrably not equally distributed, but people create a polite fiction of egalitarianism by chatting, eating, and drinking together as they work.

The networks of neighbors, kin, and friends who exchange labor and other goods and services are one of the major links between household- and parish-level social units. Yet the existence of such networks has not kept farmers from adopting labor-saving machinery. Nonfarmers often complain that farmers now use so much machinery that there is no more work in agriculture, contradicting the farmers' complaint that no one wants to work anymore. While new technology generally reduces labor needs, some tasks, such as silage cutting, require large crews, although only for a few days (see Chapter 3), reinforcing the need for large labor parties.

WATER AND IRRIGATION

The two major water sources are *poços* (wells) and *poças* (ponds). Well digging is an old technology, although much more common now than in the early 1960s. While the water table is only about eight meters below the surface of the ground, the lower six meters are usually solid granite. Formerly wells were dug by hand and lined with stone. Dynamite and air hammers have made well digging easier, and prefabricated concrete rings have made lining the well much simpler.

Persian wheels, or norias (*estanca-rios* or *noras*), used to be used to draw water from a well using a chain of buckets, powered by a cow walking around the well attached to a big cogged wheel. The Persian wheels are now rusting on the old wells (see Pinto 1983). Since 1978 all farmers use small pumps, either gasoline-powered or electric.

Around the same time local stone masons started building *tanques* (tanks) of large granite blocks in the fields of some larger farmers. Tanks at the highest point of a field are pumped full from a well. With pumps and tanks farmers can now irrigate uphill from the well; the old wells and norias had to be at the highest part of the parcel. A tank can be filled any time, quickly or slowly, to irrigate at the farmer's convenience. The tank's outflow is precisely regulated with a turn valve for a deep soaking. A slow-filling tank can be drained rapidly, giving the water enough of a head to reach all corners of the plot.

A poça is a regional type of irrigation pond, apparently old and still popular. The water source is a *nascente* (spring) (usually with some rock removed from the spring head to improve water flow) or a *mina* (horizontal well). A pond is formed inside a rectangular or rounded stone wall built against the hillside at the spring. Few ponds are more than a few meters across, and most are less than a meter deep, although one large farmer owns two the size of swimming pools. Horizontal wells are slighly inclined, hand-dug tunnels in the hillside, so the water flows by gravity from the aquifer to the pond. Ponds are drained from a sluice gate (*o olho-d'água*, lit. the eye of the water) or with an *engenho* (a stone siphon). Ponds uphill from the land they irrigate need no pumps to tap the water. Although poças were designed before machinery was readily available, they are still adaptive because, unlike the Persian wheel, the ponds are at or above the high points of the fields they serve and need no machinery. A head of water is call a *pé* (foot) in Portuguese, and water that flows by gravity is said to *correr a pé* (run on foot). Farmers who had not adopted irrigation pumps invariably explained that they had not done so because all their water ran on foot.

Unlike wells, which are almost always owned by a single household, many poças have several owner, or member (*sócio*), households that send workers (usually household members) to clean out the pond on St. Peter's Day—June 29—after which it is on a summer schedule until the day of Our Lady of Porto (Nossa Senhora de Porto, September 8), when it reverts to the winter schedule. Usually the owner of most of the summer shares also has all of the winter water. See Table 2.1 for a pond schedule, only slightly more complicated than the average.[9]

Some Portuguese ethnographers refer to cooperative and collective ownership of resources in the North, explaining them as Tylorean

Table 2.1. Irrigation Pond Schedule for A Poça de Subarribes

Summer	
Sunday	Enough water for the gardens of *A* and *B*, the remainder for *C*
Monday	All water for *C*
Tuesday	All water for *C*
Wednesday	All water for *C*
Thursday	*C* until 9:00 A.M., one pondful for *E*, one pondful for *F*
Friday	*E* first Friday, *C* second Friday, *D* third and fourth Fridays of each month
Saturday	*C* mornings, *E* afternoons one week; on alternate weeks *C* afternoons, *E* mornings

Winter	
Sunday	*C* one week, *D* one week, *E* one week
Monday	*F*
Tuesday	*F* until 9:00 A.M., *E* until sundown, *C* after sundown
Wednesday	*C*
Thursday	*C* until 9:00 A.M., then *E*
Friday	*E* until sundown, then *D*
Saturday	*D*

NOTE: Each member household is indicated by a letter; there are six member households, *A–F.*

survivals (see especially Dias 1981; Oliveira, Galhano, and Pereira 1983). Neither cooperative nor collective, ponds are "corporate resources, neither privately nor communally owned" (O'Neill 1987a: 167) whose social organization is ecologically adaptive now, not a mere holdover of Visigothic communal tenure. A household should send workers in proportion to the amount of summer water it owns. Members may use their water, give it away, or let it run downstream, without consulting the others. Although water is not sold, a person with a dry field may ask for a water turn (*turno*) from a pond member or well owner.[10] The water giver expects the receiver to return the favor later, with an unspecified amount of labor. Wealthy households get extra labor by giving away surplus water to smallholders without water rights. Those given water may be called to work many days for the water donors. Because the terrain is slightly broken, pond members have a near monopoly of water in a small area. The water takers are reluctant to refuse to work for the donors who could dry up the little field that is often the last link on the small water network. "This free water is expensive," a one-cow farmer complained in the summer twilight after yet another day of unpaid labor for the farm family with two large ponds.

The pond is owned by a small corporation, with individually owned shares. Locals claim that theoretically water rights are separate from land rights, but that in practice they are never divorced. Each water turn is linked to a certain field (or a cluster of them). The owners of a poça have land downhill from it, and people who buy land buy the water too, although written deeds are not common. Pond members, who have no formal ties besides annual cleaning and water sharing, form one of the few formal intermediate social groups between parish and household. The groups are not communitywide institutions, since most households have no pond shares while some have several.

Reflecting the ideal of the autonomous household, recalcitrant partners cannot be coerced by fellow poça members. If households A, B, and C own a pond, and A sends no one to help muck it out on St. Peter's Day, B and C can do little but complain, generally behind A's back. Because of their loose organization—no boss, no sanctions—there is an upper size limit to a pond organization. One pond, the largest by far, about one hundred meters across, taps the water of a flowing stream. It has three outlet systems, discharges into a neighboring parish, and is on a seventeen-day schedule that is interrupted by a separate Sunday schedule. Like the Alpine Swiss irrigation systems, no one could explain the whole system (Netting 1974b). Some people claimed not to understand when their own turn was. The pond was silting up, and while people were concerned about it, no one could organize a group to clean it out. This large poça is starting to be neglected because every year there are more wells.[11] People who can choose between water from their own well and a share from a corporate pond usually use the well, which is theirs to tap any time they please. Only a few wells are owned by more than one household.

MAIZE

Maize, the dominant species in the managed plant community, is intercropped with beans and squash. Potatoes and kale compete for summer field space with maize. Wine grapes surround summer maize fields that grow grass and rye in the winter. Livestock eat crop

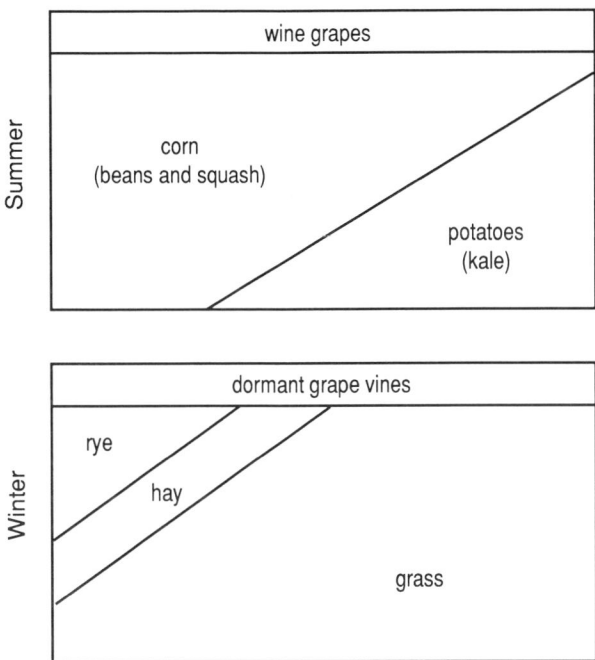

FIGURE 2.2. Schematic Time and Space Relationships of Plants in the Managed Field Community

by-products. Most households in Pedralva (including all farms) manage another plant community—a small garden. Figure 2.2 is a schematic diagram of the spatial and temporal cropping relationships.

Maize is the most important crop, occupying most of the farmer's attention and the bulk of the field space. Minhotos call maize *pão* (bread), just as villagers in Trás-os-Montes refer to their main cereal, rye, as bread (Dias 1981). Land planted in cereals is said to *andar a pão* (to walk, or rather to be, in bread).

Religious symbolism associated with cereal crops includes crosses, sometimes carved on a stone in the wall around a field. Crosses are carved on the fronts of granite corncribs, on the lintels of the old stone water-powered mills, and on the stone doors of the household ovens. When a woman makes sourdough maize bread (*broa*) she scratches a cross in the dough rising in the *masseira* (dough box). As she places the last loaf in the oven she makes the sign of the cross with the bread paddle (*pá do forno*) before the open oven door. If a

Table 2.2. Maize Seed Types and Planting Styles

	Planting Style		
Seed Types	Forage Maize	Grain Maize	Silage
Temprão (local)	X	X	
Resteva (local)	X	X	
Hybrid		X	X
Second-generation hybrid	X	X	

child walks under the bread paddle when it is in the oven, it is half-jokingly said that the child will stop growing.

Maize bread is emotionally important. It is considered bad manners to eat a meal without bread, and one cannot complain about the food if there is plenty of bread (see Dias 1981). Many people tersely summed up their life before 1964, saying that "antes da emigração não havia pão" (before emigration there was no bread). White bakers' bread is not pão but trigo (wheat).

Minhoto farmers plant maize in three major styles: milho basto[12] (forage maize, lit. dense maize); milho para grão (maize for grain); and silagem (silage). Three maize seed types—hybrid seed and two land races (local varieties, milho regional)—partially correspond to the three planting styles (see Table 2.2).

Growth cycles of hybrid maize range from ninety to 120 days. Hybrid seed, usually bought at the milk cooperative, is planted for silage and sometimes for grain. Some farmers sow second-generation "hybrid" seed (milho híbrido do segundo ano) for grain. Second-generation maize yields more than land races, although yield drops each year with its declining heterozygosity. Temprão is one of two named land races of maize; a one hundred–day variety planted early in grass fields that are cut and plowed in the early spring. Resteva[13] is an eighty-five-day variety planted later in fields that must be left in rye or in grass for hay until early summer.

Milho basto is the least common, least important planting style. Maize is sown broadcast, instead of planted in rows. As its name (dense maize) implies, it is planted close together, especially under grape arbors, in fields without irrigation, or other marginal plots. Broadcast forage maize generally forms a tassel but not ears, because it usually lacks sun, water, or both and because farmers uproot the plants in midsummer, before maturity, for cattle feed. This fills a critical gap since forage is difficult to get in summer, when most of the fields are in maize and potatoes rather than grass. Although milho basto looks like a traditional planting style, Oliveira, Galhano, and Pereira (1983) do not mention it. Informants say it was invented in the mid-1960s in response to greater demand for cattle feed.

Broadcast maize tends to be grown by farmers with larger herds. The average herd size of farmers who grow milho basto is 5.4 cows (median 3.9); those who do not raise dense maize have an average of only two cows (median 1.4). (See Table 2.3 for the distribution of herd size and broadcast maize for seventy-two farmers.) Smaller farmers grow an even more labor-intensive crop (a garden, potatoes, etc.) in marginal places. Farm households cannot be divided into two or three groups and described as typical systems. Farmers have options regarding which crops to plant and a menu of several technologies for each task. Every household uses a distinctive mix of techniques, adjusting its technology to suit its particular resources of land, labor, and capital, and adapting to its unique natural and economic environment.

Table 2.3. Relationship between Herd Size and Broadcast Forage Maize

Herd Size (including work cows)	Number of Farms Growing Milho Basto	Number of Farms Not Growing Milho Basto
1	5	15
2	5	8
3	8	1
4	9	0
5	5	3
6	4	0
7	1	1
8	1	0
9	3	0
10	1	0
27	1	0
33	1	0
Totals	44	28

Milho para grão, sometimes called *milho para pão* (maize for bread) is maize grown to maturity for bread flour (*farinha*) and for coarser meal (*farelo*) for animal feed. Grain maize used to be the main summer crop; since 1980 it has met increasing competition from maize silage.

Maize fields are manured early in April if the field can be cleared of winter grass. Some fields cannot be cleared early, since some grass must be left growing till May to cut as hay. As described above, April manuring begins by building a compost heap (moreia or *ruma*) of dung and brush in the field where it decomposes for a few weeks. Remaining fields are fertilized later, the day after the hay has been removed in May, by dropping off piles of manure (*montes de estrume*) with a *gadanha* (three-pronged rake) from the back of a wagon. This dung is not mixed with green brush because the field is plowed within a day or two of manuring, and the heat of the rotting gorse would kill the seed.

Farmers trim the field edges (*fazer as beiras*) in early spring, hoeing weeds and grass back from the edge of the stone walls, making a bare strip about a meter wide around the field. Ivy (*era*) and blackberry (silva) are peeled off the walls with hoes and piled in little mounds (*terrões*) in the field, to be dried and burned before planting. Household members manure and trim the field edges "so the field will look nice" (*para o campo ficar bonito*) rather than because they deem it profitable. Fields are planted wall to wall. Occasionally the largest farm households invest labor trimming the edges of one field while another is left for summer fallow.

Unlike Alpine communities where planting starts earlier at the lowest elevations (Netting 1981: ch. 2), in Pedralva maize is planted first in upland fields. There is little elevation range for maize fields,[14] and heavy winter rains leave the soil waterlogged in the spring. The higher fields soak up less rain and dry out sooner because the soil is thinner, being made up of more decomposed granite and less humus than the thick earth of the valley bottoms. A wet spring may delay planting (as in 1983 and 1984) up to a month in some fields. Planting dates are determined by labor demands as well as field elevation and soil moisture. Small producers plant all their fields soon after early April, while the largest farmers may not plant their last field until late June. Labor bottlenecks sometimes delay planting for over a month. Small farmers may hoe weeds from knee-high maize the same day that a larger grower plants the neighboring field.

A field can be plowed as soon as the edges have been trimmed and the compost heap has been spread out or the manure piles have been broken up. Every farmer in Pedralva plows with a tractor, including fifty-three who do not own one and twenty-nine who do. Thirty-three (73%) of the forty-five nonfarm households that grow maize plow with tractors, which can be hired by those who do not own one. Rates were Esc 800 per hour in 1983 (about U.S. $8.89 in 1983 dollars), and Esc 1,000 in 1984 (about U. S. $7.69 in 1984 dollars). Just as the same laborers work for the same farmers, Pedralvans tend to hire the same tractor owner for every task. One nonfarmer and five farmers do custom tractor work, making appointments for a certain field at an approximate time on a specific day. The tractor owner runs the machine while the client watches. After plowing the field, they may do an *acabamento* (finishing); chaining an ox plow to the back of the tractor they cut another furrow around the edge of the parcel. Alternatively, cows may plow the finishing furrow, or people may turn over the field edge with hoes, using every bit of land for maize.

After plowing, the field is harrowed to break up the clods, either by tractor or cows. Traditional harrows are rectangular and wooden, with iron spikes on one side. Some fields are flattened by rolling a stone or a hollow, water-filled cylinder over them. Oliveira, Galhano, and Pereira (1983:30–31) describe maize planting by hand, mentioning that animal-pulled planters were an innovation in the midtwentieth century, "initially heavier, and pulled by cattle, followed by lighter models generally pulled by two people."[15] The tractor-drawn maize planter is only slightly more sophisticated than those pulled by animals or people, but it is not very popular. Farmers complain that the tractor tires pack the soft earth unevenly, and many who own a tractor plant maize with an animal-drawn seeder.

In Table 2.4 I have summarized my field observations of fifteen maize plantings. While these data are not random and are biased toward horse-pulled planters,[16] they suggest that farm size (indicated by herd size) is not the determining factor in choosing a machine- or animal-drawn seeder. Six of the nine tractor owners in the sample use animal-drawn seeders because the older animal technology is more practical. Animal-drawn planters are used on one-cow farms and on the largest, with thirty-three head. Planting with cows is slower than with tractors or horses, so only small farmers—with more labor per unit of land—can afford the time to plant with cattle. Only large farmers plant with tractors. Everyone's first choice is

Table 2.4. Relationship between Choice of Maize-Planting Technology, Herd Size, Tractor Ownership, and Horse Ownership

Cow-Pulled Maize Planter

Herd Size[a]	Tractor Owner	Horse Owner
1	no	no
2	yes	no
4	yes	no

Horse-Pulled Maize Planter

Herd Size[b]	Tractor Owner	Horse Owner
1	no	no
2	no	yes
3	no	no
3	no	no
7	yes	yes
8	yes	yes
10	yes	yes
33	yes	no

Tractor-Pulled Maize Planter

Herd Size[c]	Tractor Owner	Horse Owner
4	yes	no
7	yes	no
9	no	no
9	yes	no

[a]Mean herd size is 2.2 head.
[b]Mean herd size is 8.4 head.
[c]Mean herd size is 7.2 head.

planting with horses, which move as fast as tractors but without packing the soil down unevenly. To use a horse one must either own one or be on good terms with someone who does. Personal relations affect the choice of seeding technology more than land and labor constraints do.

It is difficult to plant a relay crop of summer maize following winter rye, which is harvested around June, late for planting maize. One way to do so is called *milho entre os regos* (maize between the furrows)—planting corn in the early spring in the little drainage ditches (*margens*) between the rows of ripening rye. Maize between the furrows is a labor-intensive technique and its popularity is waning. Stooping over in the standing rye, in furrows no wider than a hoe blade, is hot, itchy work. Workers pull up the weeds in the furrows, then hoe in maize seeds there. A month later, when the rye has been cut, the maize is growing, although still stunted from the dense shade of the rye. After the rye harvest, beans are planted in the wide spaces now left between the rows of maize. Maize between the furrows is planted densely, but the width between rows makes for lower yields than those of other kinds of maize planting, although there may be more beans than in other styles of maize-bean intercropping. Although the technique has been used in the area since the early 1960s, only five households in Pedralva planted maize between the furrows in 1985.

In Table 2.5 we can see that maize between the furrows is grown on small farms. The cow-to-land ratio of the four households that have cattle is one cow to 3,614 square meters of land, considerably higher than the community average of one to forty-five hundred, suggesting that these are households with high labor applications and efficient land use. As the ratio of labor to land increases, land be-

Table 2.5. Size of Farms Planting Maize between the Furrows

Herd Size	Field Area (in sq m)
0	2,500
1	7,180
2	8,280
4	8,380
5	19,530
Mean 2.4	8,675

NOTE: Cow-to-land ratio for the four farms with cows is 3,614 square meters.

comes intensively farmed and privately owned (Netting 1969, 1976; Guillet 1981; Brown and Podolefsky 1976; Rhoades and Thompson 1975). A higher ratio of labor to land is associated with lower returns (value, or efficiency) to labor and higher returns to land (Netting 1974a, 1982a; Hanks 1972; Lipton 1964; Brown and Podolefsky 1976).

In 1982 a Pedralva farmer with a relatively dry plot invented a new style of planting. After plowing, he dug deep furrows with a potato digger, then planted maize in the furrows. After the maize had sprouted, he filled in the furrows with soil, rooting the plants deeply. Although labor-intensive, deep planting conserves moisture and keeps the roots closer to water. Because beans do not thrive when planted so deeply, he interplanted the beans in separate rows—two rows of maize, one row of beans. The farmer was pleased with the technique he has created and called it *o sistema das reguinhas* (the little-furrow system).[17] He now claims that some of his neighbors have copied the invention.

This case underscores the dynamic character of local technology and the intellectual life of farmers. Many Portuguese professionals, including agronomists, see farmers as changeless, tradition-bound, and homogeneous. At best the farmers are seen as accepting innovation from the outside, when technology really changes by invention as well as by diffusion. "Anthropologists, by emphasizing the patterning in traditional societies, have unintentionally contributed to a cumulative distortion in our image of the practice of traditional agriculture" (A. W. Johnson 1972:151). Anthropologists place so much store in culture as socially inherited knowledge that experimentation and innovation are ignored in traditional communities, much as Johnson's pioneering article on the subject was ignored until recently.

Farmers observe their farms and think about them. An introspective farmer with a keen intellect may create and test new techniques spontaneously (A. W. Johnson 1972; Richards 1985, 1986, 1989a, 1989b; Box 1988; Rhoades 1987; Rhoades and Bebbington 1988; Brammer 1980; Kerr and Posey 1984; Lightfoot 1987; Bentley 1989a, 1989b, 1990, 1991). A technique that is a better adaptation to the existing environment may be adopted by other farmers as well. A. W. Johnson (1971) concludes that small farmers in Brazil experiment but do not accept innovations when traditional techniques are more practical. Portuguese farmers are generally open to new ideas but retain traditional methods that are still the best technical alternative.

As discussed earlier, soil is enriched with forest brush (mato) and cow manure, usually at spring plowing. All farmers use chemical fertilizer on potatoes, and most apply it to maize too. Chemical fertilizer is always used with hybrid maize, but land races may be grown with natural manures alone. Chemical fertilizer is applied at maize planting and sometimes as a side dressing at the first hoeing.

Chemical fertilizer is granular, comes in fifty-kilogram bags, and can be spread before plowing with a simple mechanical fertilizer spreader, either hand-pushed or attached to the tractor. Smaller farmers apply fertilizer by hand, from a bucket. Fertilizer is always spread by hand when used as a side dressing on maize.

The three weed-control technologies in maize fields are hoeing, horse cultivation, and herbicides. Maize is hoed twice between June and August. Workers move slowly up the rows, abreast of each other, one person per row, chopping weeds from between the rows and from the base of the maize and bean plants. On smaller farms household members weed maize, but on larger farms outside labor is used.

The several small steel blades of the horse cultivator cut the weeds between the rows of maize but miss the ones at the base of the plants. One person leads the horse while another holds the cultivator's handles. The horse cultivator is faster than hand-hoeing, even though after cultivating with a horse the remaining weeds must be hand-hoed. Some use horse cultivation only for the first weeding, when the maize is still small enough for the horse to walk down the rows without breaking the plants. Farms of all sizes use horse cultivators (see Table 2.6). Average herd size is 7.1 for farmers who use horse cultivators, compared to four for the community average. However, horse cultivation does not depend on farm size but on personal preference and an available horse, sometimes borrowed. Some of the largest farmers apply preemergent herbicides at maize planting, but only in fields for silage.[18] At least one farmer who used herbicide also used a horse cultivator.

Several factors determine the choice of weeding technology. Farmers who own a horse use it to cultivate maize. Only the largest farmers, those facing the greatest labor constraint, use herbicide. There is a tendency for small farmers to weed with hoes.[19] Because horse cultivation is fast, inexpensive, and scale-neutral, owning or being able

Table 2.6. Herd Size, Tractor Ownership, and Horse Ownership of Farmers Using Horse Cultivation

Herd Size	Tractor Owner	Horse Owner
0	yes	no
2	no	yes
2	no	no
3	no	no
3	no	no
4	yes	no
6	yes	no
7	yes	yes
8	yes	yes
10	yes	yes
33	yes	no
Mean 7.1		

NOTE: Herd size includes dairy and work cattle.

to borrow a horse may induce any farmer to cultivate with one.

During the first hoeing farmers may replant maize if little sprouted in some spots on the field. During the second weeding, late in the summer, farmers again adjust for plant density by uprooting some maize. This thinning (*monda*) also helps alleviate the cattle feed shortage in July. Most winter grass fields are in maize during the summer, and maize by-products are the most common *comida verde* (green feed). Although cattle eat *comida seca* (dry feed), such as dried maize stalks or hay, they prefer green feed. Cows eat more and give more milk when given fresh feeds.

Irrigation is indispensable for growing grain maize, and land without water is planted in winter rye and left fallow in the summer, with light grazing, or is sown in broadcast forage maize. Maize is irrigated about once a week through most of July and August, depending on rainfall and access to water. Irrigating is a solitary task. A person meets the water where it enters the field and guides the liquid downhill with a hoe as slowly as possible, soaking the earth. Since maize fields are rolled flat when they are planted they have no furrows until the first irrigation. The farmer digs a furrow down one row for a few meters, then turns it down another, creating an angular path and allowing water to flow down the furrow, overflowing onto several nearby rows of maize. The slow-moving water is controlled by moving dirt clods around, opening and closing the short furrows, carefully watering each plant.

In August and September farmers take the tassels off maize plants (*tirar o pendão*) to get green fodder for cattle during the late summer. Farmers (or more likely their children) gather several armsful of maize tassel daily, breaking off the tassel at the first node above the highest ear with thumb and index finger. Taking the maize tops does not lower yield, as it is done after the ears have been pollinated. Taking the tassel shortens the plant and happens to make it less likely to blow over in a high wind, but farmers do not harvest tassel to wind-proof their fields; they take only what they need for fodder, leaving any extra in the field.

The grain maize harvest begins in September, peaks in October, and continues into November. Cattle owners cut maize stalks with sickles, just above the brace roots, haul the stalks home on a wagon pulled by cows or a tractor, and husk the ears at home. Nonfarmers without cattle who grow a small maize patch may bring the stalks home for husking, or they may husk the ears on the standing stalks. Either way, nonfarmers may give or sell the maize stalks to a farmer. One nonfarmer has a standing arrangement with a farmer in which the farmer plows the nonfarmer's field every year in exchange for the thinnings, the tassel, and the stalks (known collectively as *penso*).

Husking (*desfolhar*) is the most labor-intensive job of the maize cycle. Maize must be husked fairly soon after harvest to dry the ears well. By harvest the winter rains have begun, and maize is husked indoors on rainy days. On some larger farms husking goes on far into the night, for days on end. While large farmers must have help to husk the maize, this labor is seldom paid for or even negotiated. Everyone knows when their friends and neighbors have harvested maize, and anyone who feels inclined may drop by to help husk. Informants say that years ago the helpers arrived wearing masks, hiding their faces and giving a festive air to the husking bee (*desfolhada*). The parishioners are ambivalent about their past; while they enjoy much higher yields and have more money in the 1980s than twenty or forty years earlier, they cite the old husking bees as examples of how the older days were merrier (*mais alegre*).

Maize stalks are brought from the field in loose stacks (*gavelas*), with the ears still attached. Each worker sits on a stool, pulls up a stack of stalks, pulls back the husks, breaks the ears off the shanks, and drops them into a *cesto* (a basket, about a bushel). If many people come to help, the group will develop ad hoc specialists, someone strong to carry the thirty-five-to-forty-kilo baskets of maize up the stairs to the coberto (hay and grain barn) or the *espigueiro* (corncrib); someone to tie the empty stalks (*palha do milho*) into bundles (*molhos*) with wet rye straw; two people to tie the molhos to a pole or a tree, making a *meda*.

Before describing the meda, palha must be defined. I gloss palha as straw, although its meaning is closer to the agronomists' "crop residue." It is the parts of the plant (except the roots) that are left after removing the edible portion. Potato tops, maize stalks and husks, rye straw, bean stalks and leaves are all palha, after the tubers, ears, grain, and beans, respectively, have been removed. A meda (roughly glossed as haystack) is a stack of palha, whether of grass hay,[20] rye straw, or husked maize stalks, built around a permanent upright post (or sometimes a tree) that gives the haystack structure and stability. A maize meda is a narrow column of stalks tied to a pole. Starting at the bottom, a man[21] ties the bundles of maize stalks to the post, eventually getting high enough to need a ladder, while another person hands him the molhos on the end of a pole.

Villagers regard maize husking as a monotonous, boring task, but the large groups often produce animated dialog, with the host household providing festive meals and wine. Friends who stop by to visit husk as they chat. This arrangement may be exploited; local etiquette requires that even a person who drops by late in the afternoon be offered supper, although at the risk of being teased by others as he eats.

Poor families frequently owe farmers favors, for a ride to town, or a share of water, or a bundle of wild grass for the rabbits, or some other good or service. The debt may be repaid (or rather, a complex ongoing exchange relationship may be maintained) by sending a worker to husk maize. The very young, the elderly, and the physically handicapped help husk maize quite often, not only to make use of their labor but to enjoy the companionship and a good meal, and to feel active and useful. Informants say that in earlier years husking was a time for fun. Besides the masks mentioned above, a game was made of finding red ears of maize as they husked. There was dancing, wine, and fresh roasted sardines at midnight. Allegedly some people danced till dawn, then went straight back to work. These customs have largely disappeared now: I saw only one mask and one game of finding the *rei* (king, i.e., a red ear of corn).[22] The disappearance of these myriad partylike activities explains informants' comments on earlier times being mais alegres. Oliveira, Galhano, and Pereira consider the festive aspects of work parties[23] an essential, qualitative aspect of labor that could not be reduced to pure economic categories because labor was still bread and life, and not money or profit. The festive customs impart an almost mystical sense to the subproductive work (Oliveira, Galhano, and Pereira 1983:51). It stands to reason that some of these customs should disappear as farming becomes more market-oriented.

One reason for the recent popularity of silage cutting—and the corresponding decrease in grain maize production—is that, by substituting capital for labor, it allows farmers to contract less extrahousehold labor, especially for husking. Silage making is also a land intensification strategy, making fodder of the maize stalks and cobs. Cattle do not eat whole maize stalks and cobs, not even when green. The leaves of grain maize plants are fed to the cattle, but the stalks are composted in manure and the cobs are used as fuel. In 1983 one farmer cut some silage but harvested several hectares of grain maize. Several workers helped husk it all day. After supper he told them to go back to the threshing floor and he would meet them there to husk by moonlight. When he arrived no one was there except his wife, two sons, and myself. After complaining angrily about the deteriorating quality of local labor, he added that the following year he would grow "silage, silage, silage."

Farmers reminisce about bygone days of surplus labor, before the mid-1960s, when dozens of masked workers showed up at the threshing floor every night after harvest, to husk the ears and then dance at midnight. The fondest memory is that the workers were not paid; "se lhes dava qualquer coisa" (they would be given any little thing). The farmers complain bitterly that "agora ninguém quer trabalhar" (now no one wants to work). At first this statement puzzled me, because most Pedralvans seem to work all the waking hours. After learning something of the history of local labor relations I realized that the complaint is rhetorical: farmers resent landless community members who will no longer work for meals or for whatever farm produce the farmer decides to give them.

Different segments of this little community have opposing beliefs and morals. As Harris (1985:350–52) suggests in his comparison of caste and class, social structure seen emically from the bottom is more often judged exploitative and unjust than it is from the top. Farmers and nonfarmers have different moral values about work and pay, values conditioned by socioeconomic status. The more land one has, the more likely one is to believe that other community members should work almost for free, and that they no longer do so because they are getting spoiled, putting on airs. Farmers say that "o povo é muito fidalgo" (the people are very lordly). As one of them put it, "People only want to work eight hours a day now, but they want to eat meat!" The less land one has, the more likely one is to believe that the farmers are devious, manipulative, and stingy, and that before the 1960s they cruelly exploited the poor. These distinct opinions and values reflect the tension and antagonism between local socioeconomic groups.

Sometime after the husking the maize ears are carried to the threshing floor (eira) to dry for several days. Some maize can also be dried in the barns, which are two stories tall with wide doors on the south side, so grain can be spread over the floor, with the doors open on sunny days and closed when it rains. Dried maize is stored in the barn and the corncrib, next to the threshing floor, which is usually just south of the barn. The granite lintels of most corncribs are inscribed with the date they were built, mostly in the 1860s. The long, narrow structures have stone frames, tile roofs, and wooden slat sides, oriented to catch the prevailing winds and keep the grain dry.

Dried maize is either shelled[24] throughout the year, as needed, or is shelled all at once. Only rarely does anyone still shell maize by pounding it with a malho (threshing flail). A mechanical thresher (malhadeira, debulhadeira), attached to the power takeoff of a tractor, can shell the maize from a whole field in a few minutes, at a cost of a few hundred escudos.

Maize kernels are sometimes dried on the theshing floor again before being milled. Maize is ground into flour at one of the taverns (tabernas) or tascas (small general stores), in an electric mill. Several households have their own electric mills. Gasoline-powered mills were popular until electricity was introduced in 1978. One nonfarm household still makes its living grinding flour for others in a traditional, water-driven, stone gristmill. Another farm family still operates a water mill for grinding its own maize, but most of the water-powered mills have been abandoned. It is too much work to haul the grain to isolated canyons where the streams run fast enough to turn a wooden, turbine-type, horizontal wheel. And the stream in the par-

Table 2.7. Farm Size and Herd Size of Silage Producers and Other Farmers[a]

	Silage Producers	Percentage	Other Farmers
Number	8	0.11	65
Total farm area[b]	375,510	0.25	1,065,600
Mean farm size[b]	46,940		4,930
Total number of cows	115	0.37	194[c]
Yearly milk production[d]	209,000	0.43	277,200
Land-to-cow ratio[b]	3,265		5,492

[a]N for sample is seventy-three farmers for whom complete data were collected.
[b]Area, farm size, and land-to-cow ratio are given in square meters.
[c]This figure includes all of the parish's thirty-one work cows.
[d]Production is given in liters.

ish center is said to be drying up as a result of recent well digging.

There is more than one way to do most grain maize tasks, and choice of which to use is determined by the supply of the factors of production, especially land. However, the most important innovation in maize growing, and one whose adoption depends most clearly on factor supplies, is maize silage. The largest farmers, with the most land and capital and the least labor (per unit of land) grow most of their maize as silage, planting just enough grain maize for household use (see Table 2.7). Silage producers are only 11 percent of the farmers, yet they farm 25 percent of the cropland. They have 37 percent of the community's cattle, marketing 43 percent of the milk. They get more milk per cow because they own none of the parish's thirty-one work cows, which do not give milk, and because they have enough land to leave some in summer pasture instead of grain maize. They have less household labor per unit of land and seek to substitute silage choppers (capital) for the hand husking and processing of maize (labor).

Maize silage is a recent adaptation to labor constraints and to an increased demand for cattle fodder. Whereas grain maize is ground into bread flour and cattle and pig feed, all maize silage is fed to dairy cattle, to produce milk for sale. Silage is a new orientation away from the community toward the national economy. The technology was learned from extension agents in the late 1970s, and in 1984 was being changed to meet local conditions. As with grain maize, green manure is applied to silage, and soil preparation is similar. But silage is not intercropped with beans and squash, and hybrid seed (ninety- to 120-day varieties) is generally used, as are chemical fertilizer and ground limestone.

Before 1978 all cattle were kept in *cortes* (stalls), small, dark rooms, often on the ground floor of the farmhouse. Since 1978 the three largest farms have built modern cow sheds (*vacarias*). The manure, urine, and wash water collect in a large underground *fossa* (cesspool). The three large farms have adopted a number of ways for spreading this liquid manure (*chorume*) on fields, including spraying it through metal irrigation tubes with a high-powered pump; composting it with brush and applying it with a mechanical manure spreader; and spraying it on the fields with a tractor-pulled tanker wagon.

Maize silage tends to be planted with a preemergent herbicide to control weeds before they sprout, although the fields may be horse-cultivated or hand-weeded as well. Irrigation is like that for grain maize. Silage maize may be thinned if it has been planted too densely, but more often the emphasis is on planting just the right amount, to avoid thinning. This is because hybrid seed must be bought (and cost about Esc 100/k in 1984); because the silage producers do not need the thinnings for forage (they have last year's silage, and usually pasture); and because they have less labor to hand-thin the maize. They use land instead of labor to feed the cows over the summer.

Silage cutting permits greater control over the harvest timing. Early rains may delay the grain maize harvest for two or three days as farmers wait for the maize to dry before cutting it. But silage is cut in the early autumn, during the first few weeks of the grain maize harvest. Silage can be cut even after a rain, because the maize is to be fermented, and moisture will not induce mildew, as it will in grain maize. Silage is cut with three tractors and about a dozen workers. Although this new technology requires as many people, and more machines, than any previously known task, farmers have had little or no trouble organizing helpers and equipment. Two or three farms generally pool their machinery and labor, cutting the silage first on one farm and then the others, but extra workers are usually needed as well. For the one to three days it takes to cut a farm's silage, that farm is responsible for getting the six to nine extra workers, either by hiring them or by calling in favors.

Like grain maize fields, silage fields are planted from wall to wall. Workers with sickles begin the harvest by cutting the four outside rows of maize around the edge of the field. In fields larger than a hectare a swath is also cut down the middle. The farmers attach a one-row maize chopper to the tractor's power takeoff, and hitch a wagon to the back of the tractor. During the first pass around the field previously cut plants are hand-fed into the lumbering chopper. The whole maize plant is hacked to bits like a tossed salad and blown up a tube into the wagon. When the wagon is full, the tractor driver unhitches it and hooks up a waiting wagon. A second tractor pulls the full wagon to the horizontal silo. Most wagons now have hydraulic lifts and are easily dumped. Other workers help empty the wagon and meticulously spread each wagon load around the silo, generally an open, concrete-lined trench sunk into the ground. A third tractor slowly drives back and forth over the chopped maize, squeezing out the air so the silage will not rot. Five kilos of salt are mixed into each ton of silage. The filled silo is covered with black plastic sheeting, sandbags, and old tires, allowing the silage to ferment in an airless environment.

There is some intracommunity variation in silo construction. Early silos were built according to the extension agents' specifications: concrete-lined, about three meters deep, with high awning roofs. Later silos are shallower, without roofs, lined with cement block, and still more recent silos are simple earthen trenches, sometimes without a lining of cement block. In 1984 farmers realized they could ferment a pile of silage on the surface of the ground, covered with a plastic sheet. Local adaptations of the original technology save money and are technically sound. Although extension agents often complain that farmers are slow to follow their advice, examples like this suggest that much outside technology is not well suited to local conditions, or is needlessly expensive, so the farmers' suspicion and skepticism is warranted.

Back in the field one or two other workers pick up loose ears of maize and stalks that have fallen over. Another person usually works in the wagon, packing the silage down so the wagon will hold more. The emphasis is on productivity of land and machinery, not on labor efficiency. While labor is a scarce factor relative to the 1950s, and large farmers have relatively less labor than small farmers, overall land is still the scarcest factor in the Minho, even for the larger farmers. Stomping down the cut maize in the wagon makes it difficult to dump and it has to be dug out with rakes, but tight packing saves machinery time (capital); the tractor can run longer without stopping to switch wagons. The farmers strive to keep the chopper and tractor moving constantly around the field, methodically chewing up maize plants.

Villagers say, as maize is made into flour, "o pão dá muitas voltas" (the bread takes many turns). Unlike grain maize, with its many processing stages, maize silage is harvested and processed rapidly in a single stage. As already mentioned, this has a great appeal to the largest farmers who dislike depending on extra workers for many days. Farmers who do not like to use nonhousehold labor cut silage with as few as five workers: three tractor drivers and two others to tend the silo. The largest farmers tend to have more silage, more workers on call, and a greater desire to speed up machine use, so they employ half a dozen workers in such marginal tasks as packing and unpacking a wagon by hand. On the surface, this extravagant use of labor contradicts their desire to avoid the dependence on non-household labor necessary in maize husking. However, even when labor is employed in many marginal tasks (picking up fallen ears and hooking them in standing plants for the chopper, for example), cutting maize silage saves labor compared to grain maize processing. While large farmers have less labor than their neighbors, compared to those in more-industrialized countries all Minhoto farmers enjoy abundant labor and use it lavishly.

A large crew can easily chop a two-hectare maize field in one day. One early estimate claims that a single hectare of grain maize requires 252 hours (twenty-five days) to harvest and process (not counting transportation) (Lourenço and Alves 1968:165). Even when using abundant labor, a two-hectare field of silage takes only twelve person-days to process, compared to fifty person-days to harvest two hectares of grain maize. In 1984 an early autumn gale was strong enough to blow down a couple of grape arbors in Pedralva, and much of the parish's maize for silage. The tall silage maize caught more wind than the grain maize, which had had the tassels removed. One farm that took three days to harvest its silage in 1983 needed two weeks to cut it in 1984, because most of the maize had been knocked over and had to be hand-fed into the chopper. Unless farmers make further adjustments of silage maize to the local climate, cutting silage will not save labor every year.

BEANS AND OTHER CROPS

Beans (*feijões*) are interplanted with grain maize (but not with silage or broadcast forage maize). The bean seed is mixed with the maize kernels in the planter, for planting together. Several kinds of beans are distinguished by color, but the major distinction is between bush bean (*resteira*) and climbing bean (*feijão a subir*), which grows up the maize stalk. A small yellow bean, *feijão galego*, may be planted in late June, after the rye harvest, or may be used to replant where the beans originally sown do not sprout.

Beans are weeded and irrigated along with the maize, and are harvested in September, about a month before the maize. Small groups of women move in a line down the maize rows, picking the climbing beans and putting them in their apron pocket, pulling up the bush beans by the roots and grasping them in the front of the apron. The sharp edges of the maize leaves scratch the women's faces as they hurry down the furrows. Little fanfare is associated with this harvest and few or no extra workers are called.

The women carry the bush beans home in bundles on their heads and spread them on the threshing floor where they dry for two weeks. Climbing bean pods are also dried, then shelled by hand. Heaps of dried bush bean plants are beaten with a flail, knocking the beans from the pods so the palha can be raked off, leaving a pile of beans on the threshing floor. Some have found a new bean-threshing style: driving a tractor back and forth over the piles of bean plants instead of flailing them. Once the beans have been removed, the bean plant residues (*a palha do feijão*) are fed to the cows a handful at a time. It is said that the cows' stomachs swell if they eat too much palha. The beans are swept up off the threshing floor and cleaned in a hand-cranked grain cleaner (*limpadeira* or *tarara*). In 1984 several farmers experimented with a red bean an emigrant had brought from France.

Squash is occasionally interplanted with beans and corn, harvested a couple of weeks after the maize. Many people do not care to eat squash and feed much of the crop to animals.

A kale variety, known as *couve nabiça*, is sometimes planted by seed with maize. It grows very slowly in the shade of the maize and by autumn is a slender stalk with few leaves. In late autumn it is transplanted to the household garden. Its seed stalk (*grelo*) is eaten as a winter vegetable.

Like maize, potatoes (*batatas*) are a summer crop. Although two or three households have experimented with maize-potato intercrop-ping, kale (couve) is by far the most common plant intercropped with potatoes. Potatoes are planted in the early spring, especially in April, about a month before maize, and take about ninety days to mature. Thus land in potatoes cannot also grow maize. Even though farmers prefer the heavier, wetter soils for potatoes, maize and potatoes can be grown on any field in the parish. Because potatoes are planted early in the spring, they cannot follow winter rye or hay, as maize can. Winter grass fields must be cut early for potato planting.

One household in 1983 and 1984 grew half a hectare of potatoes[25] for market. Others grow only enough for their table and for gifts to workers, selling a few sacks in a good year. No one grows more than a few thousand square meters of potatoes, roughly four to fifteen tons of the tubers. Yields range from sixteen to fifty tons per hectare. Although potatoes are a staple food, many households that grow them do not harvest enough for their own use and have to buy more.

There are two important planting styles and two major seed types. Most households plant at least a few potatoes, and most of them buy improved seed potatoes every year, either foreign seed (especially from the Netherlands and Northern Ireland), or from northeastern Portugal's Montalegre. They generally plant half the field in new seed potatoes and half the field in last year's potatoes, *semente da casa* (seed from the house). The second-generation seed potatoes yield less but using them avoids a cash expense. The smallest, poor-est landholders plant only old seed.

The two major planting styles are by hoe and by plow, hoe being by far the most common. One person digs a furrow in the unplowed sod with a hoe. Another person follows, laying potatoes and fertilizer in the furrow. Everyone uses chemical fertilizer for potatoes, and many use manure too. A third person covers the potatoes with a little earth, followed by a fourth who puts kale seedlings between the potatoes. Then the kale seedlings are covered. The kale seedlings have been growing since midwinter, when they were planted in a *margão*, a small, square manure pile in the garden, with a little greenhouse of plastic sheeting and poles built over it. The ratio of kale plants to potato plants is governed by the household's supply of land. There is no market for kale; everyone must grow enough for her household's soup and to feed to the chickens, pigs, and rabbits. The less land a family has, the fewer potatoes it plants. The very poorest plant kale and no potatoes.

Potatoes are planted with a plow (either cow-drawn or, more commonly, tractor-drawn) in larger fields and are fertilized with manure and chemicals. A cow-drawn plow turns over a furrow in the manure-covered sod. Potatoes or kale are placed in the open furrow, which is then covered as the sod from the next furrow is flipped over on top of the seed potatoes. Potatoes and kale are placed in different furrows when planted with plows. Tractor plowing is slightly different. First the field is plowed, then the driver attaches a three-pronged plow, *charrua para abrir regos* (lit. plow to open furrows), to the tractor, which opens up three furrows at a time. Workers put kale seedlings and seed potatoes in the bottom of each furrow, along with chemical fertilizer (spread by hand from a bucket). A person with a hoe fills the furrows with soil, covering the potatoes and the roots of the kale. Several rows of potatoes alternate with one row of kale. Only twenty-five households, generally those with more land, use a plow to plant potatoes.

Farmers claim that potatoes grow better and with less chance of rot when they are grown in very wet soils that do not require irrigation. Potatoes are irrigated less frequently than maize, because they tend to be grown in wetter soil, but the technique for watering potatoes is not very different from the way maize is watered. Potatoes are sprayed several times over the summer with commercial insecticides (against *escaravelho da batata*, Colorado potato beetle),[26] and with fungicides again leaf mildew (*mela*).

Kale is harvested continuously. Leaves can be picked from the kale plants by midsummer. A woman who prepares her household's meals walks down the rows of kale, snapping off a broad, fanlike leaf from this plant and that one.

Potatoes are harvested in late summer, about ninety days after planting, although a field will be dug up earlier if there is any sign of plant disease. Potatoes planted with hoes are dug up with hoes, and potatoes planted by plow are dug up with a potato digger (*arrancador de batata*), a modified traditional moldboard plow with a broad round blade for uprooting whole plants. Many farmers who use the potato digger like to call in a work party to help with the harvest. One person drives the tractor (or leads the cow team), while another holds the handles of the potato digger. The other helpers pick the potatoes out of the newly turned earth. Many of them work barefooted, sifting the soil with their toes for hidden tubers, which are bagged in the field in old fertilizer and seed sacks and hauled to the barn in a wagon. At the barn the sacks are opened and the harvest is spread out to dry in the semidarkness, either on the floor or on racks.[27]

Grass is planted immediately after harvesting the potatoes (the next day, or Monday if the potatoes were harvested on a Saturday). The grass grows up around the kale plants, which are left in the field for another year, or two or three years if they are allowed to go to seed. Traditional maize growing in the Minho is a perfect example of sustainable agriculture: a lot of organic matter is worked into the soil each year, intercropped beans fix nitrogen, and the relay crop of grass prevents soil erosion during the rainy winter. The fact that a crop can be raised on the same plot every year, with (until recently) no chemicals, suggests that the ever evolving farming system of the Minho historically incorporated American crops (maize, beans, and potatoes) within older principles of sound farming (e.g., the importance of organic matter in soil) to develop a local agrarian environment that, while not capable of supporting all of the human population, at least gets a relatively constant or even increasing yield from the land.

WINE GRAPES: WORKING THE AIR

Minhotos have made wine perhaps since Roman times. During the Moorish occupation wine was grown only in northern Portugal, where quality wines are still produced. While the history of Minhoto wine technology is not well known, current practices are probably quite old. *Vinho verde* (lit. green wine, meaning new wine) was made in the Dão region as early as the sixteenth century, although the vinho verde demarcated zone was not established until 1908–1912 (Stanislawski 1970:149–52). Much of the Minho, including Pedralva, lies within the zone, but good vinho verde is not produced above three hundred meters' elevation. Because of its high altitude—about four hundred meters—Pedralva's wine is technically a vinho verde, but its only commercial use is as a *vin ordinaire* (jug wine), produced and marketed by a big wine cooperative in Braga, of which at least one large farmer in Pedralva is a member. Other farmers with extra wine can sell it to local households.

Minhotos distinguish two locations for crops: *o chão* (the ground) and *o ar* (the air). Ground crops include maize, potatoes, and grass. The air crops are grapes, olives, and fruit trees—the tree and vine

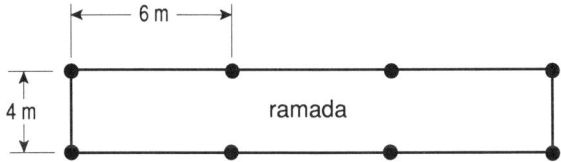

(other field, road, forest etc.)

6 m

4 m   ramada

(center of field, surrounded by *ramada*)

FIGURE 2.3. Spacing of Esteios for a Ramada

crops. A farmer occasionally rents out only the ground of a field, retaining rights to the air: "o chão é alugado, mas nós fabricamos o ar" (the ground is rented, but we work the air).

Wine grapes are the only crop to rival the symbolic importance of maize, and the only crop permanently to occupy part of the maize field. As significant as bread, wine is the favorite food and is humorously called *a pinga* (the drop). It is said that "wine is the blood of God" (*o vinho é o sangue de Deus*), and therefore "quem não gosta do vinho não ama a Deus" (one who does not like wine does not love God). When describing the misery and poverty of the 1940s and 1950s, some people said indignently that "a água era o nosso vinho [water was our wine]!" While the local brandy (*aguardente* or *bagaceira*) may be drunk after lunch or dinner, or for stomachache, or less commonly, for breakfast, no lunch, dinner, or snack is complete without a drop, for men, women, and children. Wine is believed to give strength, and so farmers offer it freely to workers. In 1983 the maize crop was excellent, the potatoes fair, but almost all the wine crop was lost. Many people judged the whole year a failure, and almost no one would allow me to see them make wine that year; they were embarrassed at how little there was.

The notion that grapevines are grown in the air is not just picturesque talk; the vines literally grow in the air. Unlike most of Europe, where grapes are often grown on short vines, grapevines in the land-hungry Minho are trained onto high *ramadas* (arbors) to free the ground below them for another crop. Esteios (usually granite posts, but increasingly made of concrete) are set out around the border of the field, in pairs (see Figure 2.3), a capital-intensive, labor-intensive adaptation to scarce land. The ramadas cradle grapes four meters off the ground, high enough for tractors, cow teams, and crops underneath. Because of the dense shade of the grape leaves, grain maize does not thrive under the ramadas, but potatoes, kale, grass, or broadcast forage maize do grow beneath the vines.

Formerly, trees—especially oak and chestnut—were planted around the field, and grapevines were trained along branches and wires strung between the trees. Vines grown this way may be over five meters from the ground, and far from the tree trunk. Some men have died after falling while pruning and harvesting grapes in the trees. Because of the danger, working in trees is considered men's work. Compared to grapes in trees, ramadas cost more to set up but take less labor to tend. Although grapes have not been trained up trees since the relative labor shortage after the mid-1960s, most of the existing trees with grapes are still tended.

Grape work begins in the winter with pruning (*poda*). Vines are trained along the wires of the ramadas. New, green vines are tied to the wire to harden there. Branches cut off to make room for new growth are tied in bundles and used as firewood.

Native grape varieties are susceptible to phylloxera (*Phylloxera vitifoliae* [Homoptera, Phylloxeridae]), a small insect that attacks the roots of the vine. Phylloxera is of North American origin and has been in Europe since the late nineteenth century. American[28] vines are resistant to the insect, so native varieties are grafted onto new-world rootstock to prevent pest damage (see Kogan 1975:105). There are two types of graft (*enxerto*). In the spring wine growers practice *garfo* (fork) grafting. A mature vine is snipped, the stump is split, and a new graft is wedged into the cleft, tied and wrapped in earth and plastic sheeting. Midsummer grafting is called *vergulha*. The bark of a mature vine is split and a new vine is spliced into it, and then bound like a fork graft.

Grafting and pruning are two of the few agricultural tasks generally done by men. This may be because, as already mentioned, the work is dangerous. Grafting and pruning are considered skilled jobs, commanding a premium wage of about Esc 600 (U.S. $4.60) per day. Female agricultural workers are not only paid less than men, they are discouraged from working in one of the few paid winter tasks, which aggravates their winter underemployment. Because of the scarcity of male workers, women have recently begun to learn to prune grapevines. When the wine harvest was ruined in 1983 villagers commented half-jokingly that it failed because women were now pruning.

Grapevines receive little attention on the ground, strengthening the notion that they are an air crop. The vines use the water and fertilizer given to the maize. In the spring farmers hoe weeds from around the base of the grapevines as part of the edge-trimming job (fazer as beiras). In the early spring the guy wires on the posts (esteios) are checked for tautness and repaired if the summer weight gain and winter loss has loosened them or tilted the posts.

The most odious viticultural task is fungicide spraying. The vines are dormant in the winter (as vides estão de lenha, the vines are in wood), but soon after they sprout leaves in early spring they must be sprayed to prevent mildew. Copper sulfate mixed with ground limestone (the Bordeaux mix) used to be the only fungicide available. The copper sulfate still comes in rock-hard lumps that have to be dissolved in water, but powdered copper sulfate and other easy to mix preparations are now available and becoming widely used. The vines are sprayed once a week, about eight times during a normal summer, less in a dry one. There are backpack and wheeled hand-pumped sprayers, and wheeled motorized ones. Twelve of the larger farms have motorized sprayers that attach to their tractors. These three hundred– to five hundred–liter sprayers are driven around the field with one person on the tractor and another walking along holding the nozzle, which shoots a powerful stream, dousing the grapes quickly from a slight distance, without copper sulfate dripping into the sprayer's face. People who use the smaller, less powerful nontractor sprayers inevitably end up with face, hands, and clothing dyed sky blue. In a humid year spraying continues until the grapes start to "paint themselves" (pintarse), that is, turn purple, after which mildew cannot spread from the leaves to the fruit.

After the grape bunches have started to mature, people break off shoots without grapes, preserving more energy for productive vines. People train the new, green vines by tying them to the arbor.

Grapes are harvested in October and should be picked within a day or two, to make the household's wine all at once. While most households make at least some wine, only the larger farmers have enough grapes to need extra labor. The farmers hire people or, more commonly, rely on their favor-exchange network to get enough workers. The vindima (grape harvest) is commonly held on a Saturday, planned about two weeks in advance, to encourage many workers.

Workers on ladders pick grapes by hand and place them in a *cesta* (a small basket, about a quarter-bushel). The cestas are emptied into cesto (bushel) baskets, which are emptied into *dornas*, large, conical, open-ended wooden casks. Wine making begins during harvest, because the wild yeast (locally called *isca*) lives in the equipment from one year to the next and begins to ferment the grapes once they are in the casks. The grapes are hauled from the field in wagons and crushed in a hand-cranked grape crusher (*pisadeira*) that is placed over the stone wine-making vat (*lagar*), dropping the crushed grapes into the vat. Small producers may make wine in a dorna instead of a lagar.

Wine ferments within two days, and is drained out of the lagar through a spout at the bottom into *cântaros* (twenty-seven–liter metal or plastic jars), then poured by hand into five hundred–liter wooden barrels (*pipas*). The wine is not filtered, but dregs may be removed later by decanting the wine as it is drained from the barrel, either before drinking or when blending with other wine. People like to have about three hundred liters of wine per household member per year. This calculation includes children, who drink less wine, but the wine is also shared with workers during festive work parties. Women pack a bottle of wine in their husband's lunches and give it to their young children with meals. Hardworking farmers sometimes drink three bottles of wine before sundown.[29] Although they consider wine one of the day's indispensable pleasures, drunkenness is frowned upon. One prosperous young farmer found one of his regular workers in a tavern in neighboring Póvoa de Lanhoso. While not exactly falling-down drunk, the big man was in a backslapping mood, in love with the world, the center of a little circle of pals. The farmer soon left and later told his father how disgusting it was to see the man so drunk. Most drink quietly without showing much drunken behavior.

Extra wine may be sold locally, although many households do not produce enough and buy more. One reason wine is highly valued (besides its fruity taste and mood-altering characteristics) is that it is the only carbohydrate food that can be preserved from one year to the next, so it is a buffer against a poor harvest of other crops. Maize and potatoes spoil by the end of one year, while wine keeps for several years. If wine spoils before it is drunk—as sometimes happens during its first year, especially if the alcohol content is low—it is sold to be distilled into the local brandy. Food storage must have been important in temperate climates where (before electricity) freezing was impossible and drying was difficult because of the winter rains.

Must is called vinho (wine) while it is liquid, but after the wine has been drained off must is called *bagaço*, related to the term *cangaço* (empty grape stem). Bagaço is squeezed in a hand-cranked wine press (*prensa do vinho*) to force out the last drop of wine. About one-seventh of the volume of the original grapes remains as pressed stems, seeds, and skins. This compressed mass, still called bagaço, is traded to local liquor distillers. The distiller comes by appointment to the house, takes apart the press, and hauls away the bagaço. The wine producer is given one liter of aguardente for each bushel of squeezings. Convention defines the bushel as the amount of crushed grapes the distiller can cram into a basket. The client leaves the press assembled as a courtesy to the distiller, so the squeezings will be in tight chunks, ideal for mounding the basket as high as possible.

The liquor merchant stores the pressed grapes in a trench like a silo and distills them in small batches until the spring. A still (*alambique*) drives off the alcohol—which is collected in copper tubing—yielding about three liters of roughly one hundred proof bagaceira per bushel of bagaço. The stills burn bark and sawdust from local sawmills. The three distillers in Pedralva are also large farmers, and the exhausted squeezings and ashes from the still are plowed into their fields, part of a long nutrient cycle of forest and field products returning to the fields after farmstead processing.

WINTER CROPS: RYE AND GRASS

There is less farm work in the winter. Farmers grow a few thousand square meters of rye (*centeio*) for the straw and the grain, but most summer maize fields are planted in grass for the winter. Rye straw is used for fodder, for tying bundles of maize, and for making a clean surface in a barnyard for slaughtering and butchering. Rye grain is mixed in the maize meal to add gluten to the bread dough, which is about three-fouths maize, one-fourth rye flour, and a dash of white flour. Larger farmers sometimes sow a mixture of rye and grass seed (*ferrão*) as a summer forage crop.

Rye fields are manured, plowed, and planted after the maize harvest, requiring little attention until they are reaped in June. Smaller farmers cut rye with a scythe (*gadanheira*), while larger farmers cut the fields with a side-mower on a tractor (*gadanheira de tractor*) or with a hand-pushed power grass cutter (*motor-gadanheira*). The cut rye is tied into sheaves (molhos), which are arranged in large stacks (*medeiros*; cf. meda, haystack). The medeiros are built on the threshing floor or on the edge of the field. In those rare cases where a field lies fallow through the summer,[30] medeiros are stacked here and there throughout the field.

Rye dries for several months in the medeiros and is threshed in August or September. Small farmers knock the grain loose by beating the rye bundles against benches or by pounding the sheaves with a threshing flail (malho), usually done by one or two household members, with none of the festive or ritual aspects described by Dias (1981) or Oliveira, Galhano, and Pereira (1983). Larger farmers hire a machine (with operator) for the rye threshing (*malhada*). The tractor-powered mechanical thresher is pulled in from a neighboring parish by a farmer who does custom work. Twelve to fourteen people are needed for a rye-threshing crew, to cut open the bundles of rye, feed them into the machine, haul away the threshed bundles of straw (*cormos*), and make the larger bundles (*cormeiros*) and the haystack. Threshing is miserable work; in an effort to keep up with the machine the workers race back and forth across the threshing floor, struggling with heavy pitchfork loads of straw, while the air is filled with a choking dust of dry rye straw, and the noise is so loud that workers must shout to talk.

A large crew is essential, and because the work is miserable people are less willing to come to a rye threshing (than to, say, a potato or wine harvest), so farmers organize a rye-threshing party more carefully than some other jobs. Several farmers make an appointment with a threshing machine operator to visit all of their farms on the same day, agreeing beforehand to cooperate with each other's threshing, often with several members from each household participating. Each farmer requests several other people to help as well, so rye threshing is labor exchange for the large farmers but part of a more complex helping (ajuda) relationship for poor people, who work for their friends (certain large farmers), in return for unspecified favors.

The mechanical thresher takes only about an hour to thresh the rye of even a large Minhoto farm. In Trás-os-Montes, where rye is the dominant crop, a large farm may take two days to thresh. In Trás-os-Montes a more complicated, villagewide network handles the mechanical threshing, and larger farmers enjoy the highest returns to pooled labor; a small farmer may work for the large farmer for two days, which the large farmer pays back in a few hours helping thresh the smaller farmer's grain (O'Neill 1981, 1982, 1987a). Most smaller farmers in Pedralva opt out of the inequity of exchanging labor for rye threshing by doing the job alone, by hand.

Two large farmers jointly own a baling machine and bale their hay and rye straw. In 1984 one medium-sized farmer had his rye straw baled with a rented hay baler. Other farmers store their straw in haystacks or in large bundles (cormeiros) in the barn.

Four varieties of grass (erva) are also grown: erva castelhana (Lolium multiflorum); azevém (Lolium multiflorum westepwoldicum); azevém verdeal (Lolium multiflorum ssp.); and erva molar (Holcus lanatus). Farmers grow grass for cattle, and nonfarmers grow it for their rabbits. Grass is a relay crop with several other plants. When potato fields are harvested, grass is planted immediately. Larger farmers sometimes harrow the seed into the ground, while smaller farmers hoe it in. After beans are harvested farmers sow grass among the standing rows of maize, hoeing in the seed. When the maize is harvested the grass is already growing in the field, preventing soil erosion, thus sustaining the land and increasing returns to it.

Grass is harvested constantly through the winter. Cows are either allowed to graze, or, especially on smaller farms, grass is cut and taken to the cattle stalls. Small farmers cut grass with a scythe; larger farmers use power grass cutters, either hand-pushed or tractor-driven. Grass is occasionally irrigated during the winter.

Erva molar and erva castelhana are the most common grasses. In the spring each farm stops cutting one grass field, or a part of one, allowing it to go to seed. The rest of the grass is cut or grazed and the field planted in maize. Erva molar and erva castelhana are cut for hay in May and June. As the grasses are allowed to mature for hay, several wild flowering annuals thrive in the fields, including rosa amarela (Chrysanthemum segetum), soagem (Echium plantegineum), and labresto (standard Portuguese saramago; Rapranus rafamistrum). Cows eat these plants as they graze, and some of these wild plants end up in the hay fodder, although workers try to pick weeds out of the grass that is dried for seed.

The cut grass is dried in the field, lying on the ground for several (it is hoped sunny) days. If it does not rain, people turn the hay over (mexer a erva) every day or so. When the hay is dry it is taken home and stored in the barn or in a haystack. Minhoto farmers are generally devout Catholics who avoid farm work on Sundays. The only time I saw a crew in a field on Sunday was when a sudden shower threatened a field of drying hay. The farmer called to me as I passed to explain apologetically that the rain was forcing them to work.

Grass that is cut specifically for seed is doubled over and tied into little bundles (molhinhos) in the field, preventing the grass from losing its seed and protecting the seed heads from moisture in the likely event they will be rained on before drying completely. The little bundles are dried in the field for a few days before being taken home to dry some more. The bundles are opened any time from June to September and pounded with rakes on the threshing floor to knock the seed from the grass. The grass is then stored as hay, and the seed is swept up and saved for the next year's planting.

Azevém and azevém verdeal are not very popular because they cannot be cut for seed until July, so they compete with summer crops. Only a few larger farmers with a relative abundance of land grow them.

The native grass "sheep's tongue," língua da ovelha (Plantagus lancelota) is a small, broad-leafed plant that grows well in wet, shady areas. Some farmers collect the seed and plant sheep's tongue along a high wall, below a terrace, or under a grape arbor. The grass requires little attention and can be harvested constantly as rabbit food.

GARDENING

Nearly every household owns or rents a vegetable garden (quintal), which, like fields, has a summer-winter cycle. Summer plants include tomatoes, onions, garlic, lettuce, green beans, carrots, and kale. Winter plants include kale, couve nabiça, and grass.

The summer garden is anticipated in midwinter, when a small seedbed rich in manure (leirão or margão) is made and onion, kale, and garlic seeds are planted to sprout seedlings to be transplanted in the spring. A miniature greenhouse, or cold frame, of plastic sheeting is built over the seedbed. Some plants, such as lettuce and carrots, are planted by seed in the spring, in a ridged seedbed called a talho. Green beans are planted by seed into a narrower ridged seedbed called a galgueira.

Farmers plant kale and potatoes in the field, while nonfarmers plant them in the garden. In midsummer more kale seed is planted, so that it is ready to transplant by October or November. The November kale is planted in solid stands, without potatoes. Winter kale is called an *horta São Miguel* (St. Michael's garden) because it is planted after the harvest of summer crops. St. Michael's Day, September 29, is symbolically the day of harvest. For example, sharecropping contracts run from one year's St. Michael's Day to the next (Caldas 1981: 211). People without fields commonly plant potatoes in the garden between rows of kale. After digging up the tubers in autumn, they plant winter grass for rabbits between the kale.

Kale is the major winter vegetable and is eaten in soup twice a day. The soup is an important source of liquid for these people, who may drink of cup of water on the hottest days as they work in the fields. It is believed that too much water can be harmful. As mentioned earlier, other plants are grown only after the household's supply of kale has been met—so the sign of a very poor family is a small garden with only kale. Some of the summer vegetables, such as lettuce and tomatoes, are perishable and are not eaten in winter. Garlic, squashes, and onions keep fairly well. Carrots and potatoes keep, but in poor condition.

Apple, pear, lemon, orange, fig, and olive trees are grown in some gardens. Except for olives, the fruit is not saved or preserved in any way but is eaten and given away as it matures. Olives are picked in midwinter and taken to a press in a nearby parish, where they are traded for oil. No one produces more than a few liters of olive oil. A very few families cure some olives to eat at home.

ANIMALS

Domesticated animals include cows, horses, sheep, goats, pigs, dogs, cats, rabbits, chickens, quail, and pigeons. Two types of cow are kept, the black-and-white Holstein or Frisian dairy cow (vaca turina) and a local work breed, barrosão. Work cows in general are known as *vacas amarelas* (yellow cows), because of their yellow-brown color. Only a handful of smaller farms still keep them. Yellow cows are still used for many jobs such as hauling, maize planting, harvesting potatoes, and plowing under grape arbors. Work cows are not generally milked.

Most farms now have at least one Holstein. Dairy cows do some light draft tasks, such as transport, but are kept mainly for their milk (for market). Calves and manure are important by-products. The cows are milked twice daily in a cooperative milking parlor (see Chapter 3 for a more detailed description). Cutting grass for cows, feeding them, and walking them to the milking parlor takes six to eight hours a day, so that having a cow is a full-time job for at least one household member, often a young woman. Milk sales are the most important cash income from agriculture.

The number of cows a household keeps depends on the farm size: the more land, the more cows (correlation coefficient of .81176). However, the more land a household has, the greater is the cattle-carrying capacity of the farm per unit of land. As I indicated on Table 2.7, the eight large silage-cutting farms require a mean of only 3,265 square meters of land per cow, compared with 5,492 square meters for other farms. This is because smaller farmers are more directly in ecological competition with their cattle (see Chapter 4). A small farm household consumes a lot of its maize crop. The family may eat all the grain, feeding the cattle primarily on maize by-products and grass. Larger farmers can devote a greater proportion of their grain to cows.

As suggested earlier, cattle are valued for their manure. Heavy, repeated manuring maintains the texture, fertility, and the flora and fauna of the soil. Minhoto farmhouses often have cow stalls on the ground floor, and the people live on the floor above. Until recently, indoor plumbing was rare. There are still some houses with a commode in a small room over the cow stalls. The human feces drop onto the floor of the stall and are composted with the cattle manure and brush. Few nutrients are wasted in the ecological cycle of the Minhoto farm.

Ten households keep a horse, used as a work animal for planting maize and cultivating. Horses may be loaned to neighbors as favors.

Eleven households keep sheep. Herd size ranges from one to nine head, with a mean of 3.6. Sheep are generally not herded with goats, although, like goats, they are allowed to graze rather than being stall-fed. Goats eat a rougher diet than sheep, so goats are tended in the forest, while sheep graze the stubble of fallow fields. Eight of the eleven sheep-owning households have farms, but only one of them, a nonfarm, also keeps goats. Sheep owners include a wide variety of households, with no one salient, defining characteristic. Sheep-owning households include nonfarms, large farms, those of solitary men, of families with fallow land and many children to tend the sheep, and those of families with average-sized farms.

Goats are kept by eight households, five nonfarm and three small farm households. Herd size ranges from one to fifty, with an average of four goats per household (excluding the herd of fifty). Blok (1981) suggests that goats are considered ritually impure in the Mediterranean, so much so that keeping them lowers a person's status. The word for billy goat (*cabrão*) is so offensive that it is hardly ever used by community members, although Minhotos are famous for their vulgar language. The billy goat is culturally offensive because it allows other males to mate with its females. When a person's spouse has been unfaithful, he is said to *levar cornos* (wear horns). To wear horns is so damaging (especially to males) that people may avoid the word *cornos* when discussing adultery, instead making a gesture with their forefingers of two curved horns rising from the top of the head.

Unlike sheep, goats are clearly associated with lower-status households. The five nonfarm goat-keeping households are all quite poor, while the three farm households include the one that keeps fifty goats and two households of bachelor farmers with no female members, who are therefore free from any possible stigma of cuckoldry and may keep goats with no symbolic danger.

The household that keeps fifty goats carries the nickname "Cabreiro" (goatherd), an indication of some identification of the people with their beasts. It is a poor household with six children, a resident son-in-law, and two grandchildren. They have access to brushland several kilometers away in another parish, where the man walks every day, herding his flock. At least two of the neighboring parishes also have a household of goatherds. Goatherding is an option for a poor family with brushland but no fields. (Goats eat gorse and the other scrub plants that grow on the forest floor and on rocky outcrops.) People may be hesitant to keep a goat unless they are already of low status or bachelors; but the goatherd's household has a reputation for honesty and hard work, and seems well liked by the community.

Hiring shepherds is perhaps now a thing of the past. Villagers regard the servants (usually small children) who were sent to herd sheep and goats before 1964 as tragic figures, with no time for school or play. One old man who was foulmouthed (even by Minhoto standards) and a bit hard to get along with was excused by one contemporary who said he had spent too much time alone in the woods with the goats as a boyhood farm servant.

Pigs are a chief status marker. Formerly only the wealthy could keep a pig and only the very wealthy could keep two. Hogs are kept in stalls in the house, often below the kitchen where food scraps and dirty dishwater can be dropped to them through a hole in the floor.

Although pigs are fed food scraps, most of their diet is kale and maize meal, so swine are in ecological competition with their owners. While some nonfarm households own hogs, pig ownership remains largely a luxury of the farmers. Of the eighty resident farm households, seventy-five (94%) keep a total of 130 pigs, for an average of 1.7 per household that keeps them. Of the 181 nonfarm households, only forty-nine (27%) keep the total sixty-eight pigs, an average of 1.4 per owner.

Many households also keep dogs, especially on farms. Dogs are generally kept on short chains, encouraging them to bark madly and lunge at passers-by. Their barking warns the farmer when someone is at the gate, so if people are working nearby they can go home to see who is in the yard. A few dogs are also used for rabbit and bird hunting in the fall.

Cats have a symbiotic ecological relationship with farmers and are tolerated in large numbers, but not as pets. They are valued for killing the rodents attracted by stored grain. Cats are not fed, although they often sit in the doorway or at the hearth and eat the fish and chicken bones that people throw them from the table. The only time I ever saw anyone touch a cat in the Minho was when a party was gathering at a farmhouse for a day's work, and a cat crossed the threshing floor dragging a huge rat. A man picked up the animal by the scruff of the neck, holding it out for his neighbors to see. He said "look at this big rat!" several times, but the normally talkative group regarded him with a stunned look, almost of horror, until he finally let the cat go, rat still in its mouth.

A few people keep pigeons, ducks, quail, and even small wild birds (especially jays), mostly as a hobby. Rabbits and chickens are the economically important small animals, distributed fairly evenly across the community, although farmers are slightly more likely to keep small animals, and in slightly larger numbers. Rabbits are kept by fifty-nine (74%) of the farm households, with an average hutch of 10.1 animals per household that keeps rabbits; while 108 (60%) of the nonfarm households keep rabbits, with an average hutch of 7.7 animals. Chickens are kept by seventy-seven (96%) of the farms and 129 (71%) of the nonfarms, with an average flock of 15.9 per farm that keeps chickens, and 9.1 per nonfarm. Rabbits and chickens are fed maize and kale. Chickens also run free and eat insects, and the rabbits are fed grass. Rabbits and chickens are kept by nonfarm as well as the farm households because they cost little to feed, they take advantage of small amounts of kale and grass, and they provide enough fresh meat for a single meal.

The contemporary agricultural technology of the Minho is complex and, as we will see in Chapter 3, rapidly changing. The complexity of the technology mirrors the social inequalities discussed in Chapter 1. Households are not equally endowed economically. The emic models present broadly ranked social categories (farmers, poor people, emigrants; or big farmers, poor farmers). My etic model suggests a socioeconomic continuum, with each household having a unique natural and economic environment and choosing a slighly different technological mix from its neighbors'. No two farms are operated with exactly the same technology.

One of the most important variables is land. All landholdings are small, but not equally small, ranging from no land to 8.5 hectares. As mentioned earlier, it is generally the case that the more land a household has, the more capital it has, but also the lowest ratio of labor to land. Each household has a particular combination of land resources, that is, different amounts of farmland (with different amounts of heavy and light soils) and forestland, in many or few parcels, distant or nearby. Each household has different capital and labor resources. Some have many able-bodied workers, others do not. Some farmers are vigorous, others ill. Socioeconomic diversity conditions technological diversity, as each household tries to use its resources efficiently. While a person's attitudes may reflect his or her social status and economic means, much cultural knowledge is shared. Most community members are familiar with the full range of technology, especially because they exchange labor with each other and learn techniques that they do not use on their own land. Thus a relatively homogeneous body of cultural knowledge is manifested in behavioral diversity, as people seek to maximize their particular combinations of land, labor, and capital, on their particular farm.

3. Technical Change in Agriculture

Cancian (1972, 1979) argues that large farmers generally adopt new technology first because of the "facilitating effect of wealth": buying a sack of fertilizer is less risky an investment to a larger farmer than it is to a smaller producer; farm technology is reasonably stable until influenced by inventions from outside the farm community.[1]

The "induced innovation" model of change in agriculture is based on relative changes in the prices of land, labor, and capital (Ruttan and Hayami 1984). Farmers adopt new technology to adapt to a changing economic environment; for example, machinery[2] may be adopted to adjust to rising labor costs (wage rates). This model explains why farmers may know about a new item for some time but adopt it only after changing prices make it profitable. Geertz's (1963) description of increasing agricultural intensification in Indonesia could be cited as an example of induced innovation: rising population and colonial exploitation created land shortages that forced people to substitute abundant labor for scarce land by switching from swidden agriculture to irrigated rice cultivation.

While essentially accepting these arguments, my thesis is that technical change is affected by each farm household's supply of land, labor, and capital. These three factors of production are usually linked: a farm with more land generates more capital but has less labor per unit of land. Nevertheless, qualitative analysis of case studies of different innovations in the same community show that land, labor, and capital cannot always be reduced to a single variable (e.g., farm size). In some cases the supply of one or two particular factors (or even something more specific, such as availability of a horse) may determine which farmers adapt and how quickly.

In Table 3.1 we can see that all of Pedralva's farms are small, but some are much smaller than others. In 1984 the average dairy farm had 6.2 hectares of forest and cropland in 13.3 parcels.[3] Since 71 percent of this land was forest and brush, the average farm had only 1.9 hectares of cropland, divided into 5.7 parcels. Mean parcel size for cropland was 0.33 hectares. The twenty-two farms with five or more cows owned 66 percent of the dairy animals, farmed about one-half of the cropland, and had over half of the tractors.

Farmers are community members, with a shared culture, each in a different economic environment, who respond rationally to changes in prices, nature, technology, and off-farm employment opportunities. Many behavioral differences between large and small farmers can be accounted for by household economic differences. Although farms with more land tend to house more people, they do not have as many workers per unit of land as the smaller farms. So smaller farmers are faced with greater constraints of land and capital, while larger farmers have greater labor constraints. A small farmer must use precious land and capital conservatively, compensating with plentiful household labor. Thus the small farmer may not buy a new labor-saving tool because she does not need it, not just because she cannot afford it.

Pedralva entered the local and international wage labor market in the mid-1960s, when France opened its borders to Portuguese laborers. Before then the community was divided into two social groups. Those with enough land to pasture two cows were called lavradores (farmers; lit. plowers). The landless group was named the pobres (poor). The farmers earned their living by unrelenting toil on their one- to six-hectare farms. The poor people scraped a living together by working their own small gardens, spinning textiles[4] at home, gathering chestnuts to eat and acorns to sell as hog food, and collecting firewood; a few begged or prostituted themselves, and all worked some months as agricultural day laborers (jornaleiros) at very low wages, sometimes only for their meals.

Before 1964 it would have been easier to describe labor relations in Pedralva using fewer concepts of formal economics, much as one would describe the Kula ring, the Kwakiutl potlatch, or the East African cattle complex. Many workers rarely earned cash. Some poor people sharecropped parts of the farms of wealthier farmers. The poor gave children to farmers as servants, so they could work the fields for a place to sleep in the barn, their bread and soup, and a new

Table 3.1. Characteristics and Distribution of Parish Landholdings by Number of Cows, 1984

Cows per Farm	Number of Holdings	Total Milk Cows	Mean Number of Milk Cows per Farm	Total Work Cows	Mean Number of Work Cows per Farm
0	79	0	0	0	0
1	20	17	0.85	3	0.15
2	13	24	1.85	2	0.15
3	9	20	2.22	7	0.78
4	9	34	3.78	2	0.22
5–9	10	96	5.33	17	0.94
10–33	4	87	21.75	0	0
Totals for all landholdings[a]	152	278	1.83	31	0.20
Totals for farms[b]	73	278	3.81	31	0.42

[a] A landholding is a household with land, whether rented or owned. It includes farms as well as smaller holdings.

[b] Farms are defined as landholdings with at least one cow. Eight farms were omitted because of lack of data.

[c] All land figures are in hectares.

shirt once a year.[5] One emigrant, home for vacation, told me how he despised farmers because as a child he had been a servant. One of his tasks was to lead the oxen back and forth across the fields while the farmer handled the plow. The old stumps of maize stalks are hard and sharp since maize is cut off just above the ground with a sickle. The stumps cut the boy's bare feet until they bled, but he was never given shoes.

Even those who did earn cash—the day laborers—got little money, receiving the bulk of their pay as meals and gifts of food. The social relationships between farmers and workers lasted over a long term, often a lifetime, and passed on to their children. They were highly personalized economic relations, expressed in the idiom of friendship. The poor did not sell their labor to the highest bidder but worked for the same few farmers (their friends) year after year.

Sharecropper relations were also durable and personalized. Technically, they were based on annual verbal rental contracts beginning

Total Cropland[c]	Mean Cropland per Farm[c]	Total Number of Parcels	Mean Number of Plots per Farm	Mean Number of Tractors per Farm
15.89	0.20	126	1.60	0.04
14.44	0.72	73	3.65	0
13.17	1.01	58	4.46	0.15
13.86	1.54	37	4.11	0.33
22.93	2.55	58	6.44	0.89
54.14	3.01	149	8.28	0.67
25.57	6.39	51	12.75	1.50
167.81	1.10	552	3.63	0.22
151.84	2.08	426	5.84	0.42

on St. Michael's Day, September 29 (Caldas 1981:211), but in practice they were often for a lifetime. Sometimes a young couple became tenants on the same farm where the parents of one of them had been sharecroppers, inheriting their position. Tenant and landlord agreed on a payment in kind, sometimes fixed amounts of grain and wine, sometimes a share (such as half of the grain and wine, or half of the grain and a third of the wine). I knew of one couple who were sharecroppers for a large farmer until the 1960s, when they became too old to manage the farm. Being childless, they had no one to inherit their tenancy. Their previous landlady's grown children bought a tractor and other machinery and took over the sharecroppers' farm as well as the one they themselves already cultivated but allowed the tenants to stay in the house where they had lived for years, hiring them to work as day laborers. By 1984 the old man was dead, and the woman was still living in her old house, supported by the children of her former landlady and still doing occasional field

Table 3.2. Chronology of Technological Change and Emigration Experience

	1940 1950	1960 1970	1975 1980 1984
	End of Brazilian Emigration	Emigration to France Begins	Emigration to France Declines
Labor-Saving Devices and Practices	corn shellers grain cleaners water pumps	rye threshers tractors tractor-driven corn shellers fertilizer, replacing brush flax abandoned grape crushers	silage hay balers manure spreaders
Land-Saving Devices and Practices	some chemical fertilizer and hybrid corn seed used increased number of grape ramadas	increased fertilizer use hybrid corn used on some farms	limestone commercial dairying modern potato production increased hybrid corn use

work for them "as a favor" when they were short-handed.

Not all landowners were so kind to their sharecroppers. Parishioners still tell what they consider a very funny story about a wealthy farmer in the midtwentieth century who had a farm that he always let to a sharecropper.[6] The soil was poor, and the farm was not as productive as its large size suggested, so the owner enticed tenants to let it at a fixed rent that seemed low. Sharecroppers soon tired of paying most of their harvest as rent and would leave after a year or two. Fame of the old man's meanness and deception spread so widely that finally he could find no tenant in any nearby parishes and went to Trás-os-Montes for one. This tenant dug deep furrows and planted rye between them on the high, narrow ridges, expecting the winter snow of his homeland, so the story goes. The rye was a total failure and the tenant slipped away without planting summer maize, making the owner a laughingstock whose greed had cost him a year's rent.

It would be misleading, however, to argue that there was no labor market before 1964 in Pedralva. Although the marginal product of agricultural labor was close to zero, forcing the price of labor so low that workers were given little more than their daily meals, the near-absence of cash payments is an example of hyperlow wages in a labor market, not necessarily evidence that there was no labor market. A woman's wages for a week bought cornmeal for one week's bread, yet for much of the year there was no farm work for females. The custom of giving children as servants[7] to farmers survived into the 1960s in the Minho because landless parents were not always able to support their offspring. Now Pedralvans refer to those earlier years as a time of great misery.

Emigration has been a major factor in social change and agricultural development in Pedralva and the Minho (see Table 3.2). Estimates of emigration and population changes in the parish from 1920 to 1980 are given in Table 3.3. Emigration peaked in the 1960s, lowering the local population. In 1984, 164 people (from 138 of the 261 resident households) were emigrants or had been and had returned to the parish. One hundred thirty-six men emigrated, but only twenty-eight women. Men were gone for an average of 11.6 years, women for 8.7. All of the women who emigrated joined husbands already abroad. Although emigration was heaviest after 1964, by the

Table 3.3. Parish Population, Births, Deaths, and Emigration per Decade, 1920–1980

Decade	Population[a]	Births	Deaths	Emigration[b]
1920–1929	755	278	158	8
1930–1939	867	318	144	69
1940–1949	972	333	165	42
1950–1959	1,098	367	176	103
1960–1969	1,186	375	135	402
1970–1979	1,024	326	124	116
1980–1989	1,110	—	—	—

[a]Population is resident parish population at the beginning of the decade.
[b]The number of emigrants for each decade was calculated by subtracting the population at the end of the decade from that at its beginning, adding the number of births and subtracting the deaths for the ten years.

mid-1970s France had essentially closed its borders to new male workers, allowing only "familial immigration" of wives and children joining men already in France (Brettell 1979:16).

With the massive emigration after the mid-1960s social relationships changed dramatically. Provided with the opportunity to get out, many people left sharecropping, day labor, and servitude. One emigrant told me that he was the son of a sharecropper from a nearby parish. His father had once rented a farm for a fixed amount of grain. This was more ambitious than renting it for a share (usually half of the harvest), since he hoped to produce a lot, keeping everything above the fixed rent. Although bad weather ruined all the crops that year the landowner took the tenant to court, forcing him to pay the rent anyway. The sharecropper stayed on the farm in a kind of debt-bondage for ten bitter years to pay off this obligation. His son found the life of a French construction worker incomparably better than that of a Portuguese caseiro.

After the mid-1960s laborers could no longer be hired for bread and soup (see Table 3.4).[8] But just as emigration drained off much of the labor pool, it pumped unheard-of sums of money into the parish as emigrants sent remittances home, returned with saved wages, and hired local men[9] to build them large country houses, essentially retirement homes. What better way for the sharecropper's son who left Portugal with all his belongings in a shoe box to show his success than to become an employer, hiring others to build him and his family a French-style suburban house in his native village? There were at least seventy-one new houses built between 1960 and 1980 in Pedralva.[10] Two to six men take about two years to build a large house. The job is labor-intensive, as workers build the supporting beams of cement and steel on-site, and fashion their own wooden tool handles. During this period at least eighty other houses were "reconstructed," gutted and rebuilt using only the massive granite outer walls.

Farmers began replacing scarce labor with tractors in the 1960s. They like the tractor's speed and remark that a field that took a day to plow with cattle takes an hour with a tractor and that tractors let them farm more land, without sharecroppers, servants, and jornaleiros. While thirty-three farmers used tractors in 1970, seventy-three did by 1980. By 1984 all eighty-two farms in the parish used tractors for at least some tasks. This was a dramatic change, since the first tractor appeared in the parish around 1960. Tractors were

Table 3.4. Agricultural Wages for Men in Northern Portugal (Escudos per Day)

Year	Nominal Wage	Wage in 1975 Escudos[a]
1965	34.3	94.0
1966	35.5	103.1
1967	43.5	107.7
1968	51.5	120.3
1969	59.1	126.8
1970	67.9	136.9
1971	75.4	135.9
1972	82.6	134.5
1973	93.8	135.4
1974	118.9	136.8
1975	139.3	139.3
1976	161.4	136.6
1977	188.4	125.2
1981	471.5	147.7

SOURCE: Portugal, Instituto Nacional de Estatística, compiled by Eric Monke (pers. comm.) from *Estatísticas Agrícolas* (various years) Lisbon: INE. See also Monke (1987a) for more wage data.
[a]Adjusted for inflation, using 1975 as the base year.

adopted first by the larger farmers, who faced greater labor constraints and had more money than the smaller farmers.

While only twenty-seven farmers (and three nonfarmers) owned tractors in 1984 (thirty-four tractors), their adoption has been aided by custom plowing services. Six tractors with drivers can be hired at affordable rates—Esc 1,000[11] an hour—to plow, shell corn, and haul manure and other heavy loads. Tractor-owning farmers cultivated an average of three hectares of cropland, compared with 1.2 hectares for farmers without tractors.

Some farmers emigrated and returned with saved wages. Unlike the landless, farmers invested their money in labor-saving machinery—especially tractors—rather than in new houses. This is partly because most farmers already had large houses, and partly because they were very concerned about adapting to the loss of labor. The identity as lavrador is satisfying and meaningful to farmers, and they strive to stay in farming. Just as the emigrant houses became new status symbols, by 1984 tractor ownership had become an important symbol of wealth, the independent farm life, and manliness.[12] Several men asked me to photograph them sitting regally at the steering wheel, urging me to get the whole tractor and the wagon in the picture.

Tractor adoption was a capital-intensive solution to a labor shortage, but large and small farmers reacted to it in different ways. Farmers with over two hectares of cropland generally bought machinery. They now use their tractors for everything the yellow cows used to do—plowing, pulling wagons, harrowing, planting, and other farm work. Smaller farmers adopted tractors too, but with less money and more household labor to spend they tended to rent tractors, especially for plowing. For lighter tasks they use some of the remaining work cows, or trained dairy cattle.

The move into commercial dairy production by large and small farmers is an example of successful directed change, brought about largely by government programs beginning in the late 1970s. In 1978 electricity was introduced to Pedralva, and the milk cooperative in Braga extended service to the parish. Agronomists from the Braga cooperative had previously established good rapport with one of the largest farm households in Pedralva.[13] In 1978 they persuaded this household to buy a milking machine and a cold-storage tank—at an 80 percent discount—setting up a private dairy barn. Within a few months the extension agents had organized three cooperative milking parlors in Pedralva, and a fourth in 1981, all part of the Braga cooperative. The cooperative milking parlors are joint ventures between the Braga cooperative and individual farmers. Local people provided the buildings, usually newly built, on their own land, which they rent to the cooperative. The Braga cooperative provided the milking machines and cold-storage tanks (at no charge) and hired a manager, in every case a young female member of the household that owns the parlor.

Seventy-two of the eighty-two farms got milk cows between 1978 and 1984. The thirty-six farmers who had Holsteins before 1978 kept them mostly for milk for their families. By 1979, the year after the first milking parlors were set up, twenty-seven other farmers had dairy cows. Nearly all farmers with five cows or more got dairy cattle by 1980.

Once in the early morning and again in the late afternoon a member of each farm household (usually a woman or a child) herds the cows to the parlor to be milked. As there are only four milking stalls, people line up their black-and-white cows and chat as they wait, the only leisure that many of them have during the day. The users hook their own cows up to the milking machines and the manager writes down the liters of milk taken from every herd twice daily. Milk is collected every three to four days in cold-tank trucks from the cooperative in Braga. The four farmers who own a cooperative milking parlor go to Braga twice a month and collect cash for the milk. The money is given to the owners when they come to milk the cows.[14]

Extension agents from the Braga cooperative worked closely with a few farmers in the parish to establish milking parlors. Most farmers did not meet extension agents but simply observed their neighbors, discussed dairying among themselves, and decided to get milk cows and join the cooperative (the only legal way to sell milk). National farm and consumer subsidies made milk profitable to produce (Carvalho, Barros, and Rocha 1982).

Work cows were being sold off at this time, and dairy cows neatly filled their niche, needing similar care and feed and providing manure. As mentioned above, by this time (late 1970s) most large farmers owned tractors and small farmers rented them. But renting a tractor was too much trouble[15] for such light tasks as harrowing. Small farmers experimented with Holsteins and found they could be trained to pull wagons and harrows. The agronomists and extension agents were appalled to see dairy cows in the yoke because working would lower their milk yield, but local farm ecologies are complex and unique, and outside technology must be adapted to local conditions by the farmers themselves (Richards 1985:86). The need for traction outweighed the slightly lower milk yields.

More cattle were kept after commercial dairying was adopted, so there was more manure—all of which was plowed into the fields. Essentially everyone had experimented with small amounts of chemical fertilizer[16] by the 1950s, and during the late 1970s they began using more of it. As Pedralvans used more manure and chemical fertilizer they needed less brush. Gathering brush from the forest floor for stall bedding and green manure (applied directly to the fields) was much more common before the mid-1960s (Caldas 1981:204). Brush used to be bought in the forest, and workers were hired to cut it. Now farmers dream of the days when the landless used to line up before dawn to cut brush for bread and wine, and those who want brush now usually have to cut it themselves.[17] Brush has become a free good as forest owners who wish to clear out their forest floor (to avoid forest fire hazards) offer it to anyone who will cut it and haul it away, but usually no one will, especially since those who need brush generally have their own forestland. Inasmuch as chemical fertilizer in the Minho replaced labor-intensive brush cutting, chemical fertilizer was a labor-saving as well as a land-saving innovation (see also Silva 1983).

In the dry summers gorse and other forest floor plants burn ferociously when ignited, often destroying the trees too. Because less brush is cut than in the mid-1960s, it now grows thickly in the woods, and fires are as common in the summer as wildflowers in the spring. The owners of the private forest tracts are generally helpless to put out full-blown fires, which spread from one plot to the next, destroying large tracts of valuable timber. The thick brush left standing in the forest is a strange sight to older people and has become a symbol of how "no one wants to work anymore." Almost all villagers blame the fires on arson. Some say the fires are started by *malucos* (fools, crazy people), or by hunters wishing to clear out gorse so grass will grow for wild rabbits, while others say they are started by poor people out of resentment against the farmers.[18] No one suggests that the flammable gorse is accidentally touched off by the cigarettes of passing herders, loggers, and firewood gatherers, nor do they link the rising supply of gorse to the declining supply of labor and increasing competition from manure and chemical fertilizer.

Unlike tractors and commercial dairying, some innovations are adopted slowly. Although hybrid maize was adopted by some as early as 1955, still only four[19] farmers in Pedralva had tried it by 1974. By 1984 forty-five of the eighty-two farmers had not yet experimented with hybrid maize, which is mainly for silage. Farmers have been slow to adopt hybrid maize because they say it is not much more profitable than land races, yet hybrids require more money for seed[20] and fertilizer.

Ground limestone powder was introduced[21] to Pedralva in 1974 to lower soil acidity and give greater yields of hybrid maize and other forage crops. The native maize varieties grow well in the Minho's acidic soil, with home-grown seed, less fertilizer, and no limestone, yielding about three-fourths as much as hybrids. In Table 3.5 we see that while eighteen dairy farmers (22%) had tried limestone by 1984, thirteen of them were larger producers—who have money to gamble on fertilizer, hybrid seed, and limestone. The limestone experimenters cultivate 37 percent of the total cropland.

Table 3.5. Use of Limestone on Large and Small Farms

	Large (5 or More Cows)	Small (Under 5 Cows)
Use limestone	13 (54%)	5 (09%)
Do not use limestone	11 (46%)	53 (91%)
Total	24	58 (

Because there is a wide, evenly spread range in farm size,[22] there is not a "progressive" group of large farmers and a group of "traditional, conservative" small farmers. Innovations spread to those who can benefit from them. The more capital-intensive an innovation is, the fewer farmers who adopt it. Although most adopted dairy production, and many tried hybrid maize seed, only a few made silage. In the early 1980s farmers responded to market incentives, accepted the agronomists' suggestion, and built silos and made silage. By 1984 eight farmers had adopted silage, including seven with five cows or more. So few adopted silage in part because the silos and silage choppers are expensive, but especially because only large farmers needed silage. Farmers who cut silage avoid hiring outside workers to husk maize, substituting capital for labor. Smaller farmers have enough household labor to husk all their maize, so they are not interested in silage.

Larger farmers do not have the same farm ecology as their smaller colleagues. There are different degrees of land scarcity within some communities (Barlett 1982). Smaller farmers raise maize for bread, and only the leaves and husks are fed to cattle. The cobs are used as fuel, and the stalks are composted in the stall bedding and used as fertilizer the following spring. Larger farmers grow enough maize for their table on a portion of their land, and the rest of the crop can be chopped into silage—leaves, stalk, grain, cob, and tassel. Smaller farmers have much higher ratios of land per cow than larger owners (see Table 3.6), suggesting that smaller farmers compete with their cows for grain while larger farmers do not. In a sense, maize silage is also a land intensification strategy, since it allows stalks and cobs to be fed to animals, creating more feed per hectare (although cows and not people eat the produce).

Like silage, one local invention was also adopted mainly by larger

Table 3.6. Land-to-Cow Index

	Large Farms (5 or More Cows)	Small Farms (1–4 Cows)	Nonfarms (No Cows)
Number	22[a]	51[a]	79[a]
Mean cropland[b]	3.6230	1.2625	0.2000
Mean land rented in[b]	0.3420	0.3000	0.0800
Mean number of plots	9.1	4.4	1.6
Mean herd size	9.1	2.1	—
Cropland per cow[b]	0.3980	0.6010	—

[a]Nine farms are excluded here because of incomplete land data; 101 households (nonfarms) are excluded because they neither own nor rent in cropland.
[b]Land areas are expressed as hectares.

farmers; because there were no capital expenses, the supply of land was clearly the determining factor. In the mid-1970s farmers experimented with a planting style called dense maize (see Chapter 2), broadcasting seed densely—usually in shady or dry ground—then uprooting it for fodder in the early summer, after the winter grass supply is exhausted and before the grain maize is ripe. Only forty-four farmers had adopted this technique by 1984, including all those with eight cows or more (see Table 2.3). Again, the smaller farmers needed all their land for bread.

Land supply was also the determining factor in adoption of new potato technology. In response to higher prices for potatoes in the mid-1970s, farmers began planting improved seed potatoes, with more fertilizer, in larger, more open stands. This was largely the initiative of a Pedralva farmer who went to northeast Portugal's Trás-os-Montes potato-growing region to learn the technique from other farmers. Before the mid-1970s potatoes had been intercropped with garden vegetables, in small plots, especially in the shade. Local seed potatoes were planted with a hoe, using little or no chemical fertilizer.

Potato fields larger than one or two thousand square meters are harvested with a potato digger (arrancador de batata), a locally made modified ox plow, pulled by cattle or a tractor. The first potato digger was used in the parish in 1970. By 1984 twenty-four farmers had adopted it, including sixteen of the larger than average farmers. Smaller farmers did not adopt it because they intercrop kale with potatoes, and the unwieldy potato digger breaks a lot of kale plants.

To get the maximum yield of kale it should be widely spaced, and the shorter potato plants fit conveniently between the kale plants. Only farmers who have met their household's demand for kale on another plot grow a potato patch without kale. Potatoes are now grown to sell and are plentiful enough to be considered a staple. Previously they were reserved for Sunday and special meals.

Insecticide adoption is, to a certain point, a counterexample of the importance of the supply of land, labor, or capital. Insecticides were accepted in Pedralva in response to a change in the natural environment: a new pest. In this case, at least, chemical pest control is not the substitution of capital for land, but additional capital expended to try and maintain previous yields. Originally I did not include insecticide adoption in my quantitative analysis because I thought my data were too poor; few farmers seemed to remember when they started using insecticide. Of eighty-one resident farmers, thirty-eight said they had used insecticides "always" or "for many years." Most others said they had used insecticides for ten to fifteen years. Some locals said the Colorado potato beetle came to Pedralva around 1970. A new, exotic pest on a traditional but nonnative crop, the beetle would have had few natural enemies and crop losses must have been significant. Most farmers responded to this new pest with pesticides. Only one farmer claimed never to use insecticides and another claimed hardly ever to use them. Another eight had adopted insecticides since 1973. These ten who did not adopt insecticides or adopted them later are all small farmers, with a mean of only .7087 ha of cropland. So even in a case of widespread behavioral change in the face of environmental change, capital constraints may cause a few smaller farmers to change later than others.

Each of the innovations discussed in this chapter altered the farming system, affecting the usefulness of older techniques and other innovations. Hybrid maize became more attractive after limestone and chemical fertilizer increased its yield. As rising labor costs made grain maize more difficult to harvest and husk, commercial dairying gave maize a profitable use: silage. Dairy cows might not have been as rapidly adopted if tractors had not been replacing work cows at that historical moment, leaving empty stalls, uneaten forage, and much community knowledge about cattle rearing.

Change has been initiated by the private and public sectors. Equipment dealers, agricultural supply stores, and other farmers have introduced new tools, agrochemicals, and new breeds of plants and

animals. Government subsidies have encouraged commercial dairying, facilitated by the semipublic cooperatives (Carvalho, Barros, and
Rocha 1982:100–19). The Braga cooperative sells new machines and
inputs for potatoes, wine, maize, and rye, not just dairy. Extension
agents[23] drive to the villages to meet farmers, although the large areas
they cover do not permit them to visit more than a few of the farmers
along their beats. Larger farmers are more aggressive and confident,
and go to town to consult with extension agents. Smaller farmers
learn from larger farmers, often indirectly. I once observed a farm
worker advise a farmer on silage cutting, explaining what he had
learned working for another large farmer. The Ministry of Agriculture sponsors an annual farm machinery fair in Braga where farmers
can see and buy new machines.

Biggs (1989) argues convincingly for a "multiple source of innovation model" for generating and extending agricultural technology.
There is no one source or path of new technology, and the informal research and development of farmers themselves is considered as valid
as the work of international and national scientific research centers:
"political, economic and institutional factors have a major influence
on processes of technology generation and diffusion" (p. 12).

In Pedralva the first farm to adopt nearly all new technology was
the same household that built the first (private) dairy barn. Innovations generally spread from this farm to other large farmers and to
smaller farmers, depending on which farms can use the idea. This
diffusion reflects the close ties of communication within this densely
populated rural community. When a farmer brings home a new
machine or input, it is observed by the neighbors, so adopting new
technology is not risky for late adopters who can study it for years
before using it.

But information flow does not explain who eventually adopts new
technology. Because of differences in household supplies of capital,
land, and labor, an innovation that appeals to one farmer may be
useless to his neighbor. Villagers can save on capital expenses by
renting machinery rather than buying it, and by further adjustments
such as using dairy cows for light traction. Only larger farmers with
the greatest labor constraints get such capital-intensive, expensive
equipment as silos and silage choppers. Even innovations without
cash expenses, such as dense maize for cattle feed, may be limited to
farmers with enough land to be out of direct ecological competition
with their livestock for food.

Most innovations, including all the more important ones—tractors, chemical fertilizer, commercial dairy production, new potato technology, hybrid maize, limestone—come from outside the community, in spite of the flaws of public and private sector formal research and extension in providing appropriate technology for small farmers.[24] Local inventions include some new planting styles and adaptations of new technology, like training Holsteins to the yoke. Perhaps the pace of local invention has been quickened by the introduction of many outside innovations that served as inspiration for additional, local creativity.

4. Land Fragmentation

My interest in land fragmentation began in 1983, when I was a member of a team of economists and anthropologists studying the farm economy of five parishes in the Minho (Pearson et al. 1987). Six young Portuguese agronomists helped us interview two hundred farmers. One of our questions was, "What is your opinion of land consolidation?" The farmers were puzzled by the concept of consolidation, so the agronomists would explain that "it's putting all of your land in one place, so that you won't have to go to so many different fields. Wouldn't you like to have all of your land in one place?" The farmers either shrugged, or they argued vehemently that land consolidation would only help the rich.

I have suggested elsewhere that land fragmentation has many advantages ignored by agricultural policymakers (Bentley 1987a). Pedralva's land structure is quite fragmented and refutes two major assumptions about land fragmentation: that it is extremely detrimental to farm production; and that its main cause is inheritance divisions.

IMPORTANCE OF LAND FRAGMENTATION

Land fragmentation is the division of a single farm into several separate, distinct parcels[1] (Binns 1950:5). Policymakers and most geographers and economists regard land fragmentation as the greatest single detriment to European agriculture (Binns 1950; Dovring 1965; Grigg 1983; Jacoby 1971; Karouzis 1971; King and Burton 1982, 1983; Meliczek 1973; Naylon 1959; OECD 1969; O'Flanagan 1980; Pina-Cabral 1986:17; Smith 1978; Thompson 1963; von Dietz 1956) because it allegedly increases the farmer's travel costs as workers and equipment must move to many far-flung fields, and because the farmer wastes time maneuvering machinery in many irregular field corners and edges. A farm of a single, rectangular field would have a much larger ratio of field area to field edges than would a farm of the same size with many small, wedge-shaped, and curvilinear fields (Chisholm 1979: ch. 3).

Some claim that land fragmentation is caused by preserving outdated medieval field systems (Clout 1972b:24, 102; Smith 1978:91; OECD 1972:23; Dovring 1965:52; Meliczek 1973). Small fields were adapted to the use of animal traction and abundant human labor but are not well suited to tractors and other agricultural machines.[2]

Many critics claim that partible inheritance (especially with a growing population) leads to ever diminishing field size (Clout 1972b:41; OECD 1964:23; Moore 1972:105; Lambert 1963:31; Burton and King 1982; Brettell 1986:184). They claim that the desire to leave each heir an equal portion of the farm leads farmers to divide each field in each generation (Binns 1950). This simplistic argument is countered by an example from the Spanish Basque country. In spite of impartible inheritance practices and a positive cultural value for independent, isolated farm households, Douglass (1975) found that only some communities (e.g., Echalar) have isolated farmsteads on unfragmented holdings. In Murélaga farms are fragmented and the people live in villages. These different land structures are attributed to local differences in historical wars and original patterns of land clearing rather than to differences in inheritance customs.

Some anthropologists have discussed land fragmentation without reference to its agronomic qualities, paying more attention to its role in kinship and inheritance. For example, Herzfeld (1980) claims that dividing fields at inheritance alleviates fraternal disputes in a Greek village. But most anthropologists and geographers who have studied land fragmentation have used an ecological perspective, suggesting that, contrary to the received wisdom, land fragmentation is an adaptation to the local agrarian environment (Carlyle 1983; Cole and Wolf 1974; DeLisle 1982; Downing 1977; Edwards 1978; Farmer 1960; Forbes 1976; Friedl 1974; Galt 1979; Ilbery 1984; R. T. Jackson 1970; Leach 1968; Netting 1972, 1981; Rhoades and Thompson 1975; Weinberg 1972).

Essentially the land fragmentation debate centers on the question of whether it is an ecological adaptation or anachronistic, if not irrational and uneconomic, peasant behavior. Here, I build on the ecological argument with case studies of inheritance and a quanitative analysis of farm production and land fragmentation. Before discussing the advantages of land fragmentation, I want to consider it as a correlate of intensive agriculture and private land tenure, suggesting that the cause of fragmentation is the intensive farmers' effort to make the most of their scarcest resource: land.

INTENSIVE AGRICULTURE, PRIVATE LAND TENURE, AND FRAGMENTATION

Intensive agriculture is correlated with individual (as opposed to communal) land tenure (Netting 1974a, 1982a). Neighboring communities with different access to land often develop different tenure patterns. In Nigeria, New Guinea, and Peru, where land is abundant, it tends to be worked under shifting cultivation and is communally owned. Where land is scarce it tends to be intensively farmed and privately owned (Netting 1979; Brown and Podolefsky 1976; Guillet 1981). In the Swiss mountain village of Törbel extensively used tracts (forest, alp) are owned communally, while intensively used plots (fields, meadows, gardens, vineyards) are individually owned (Netting 1976).

The most intensive land use is correlated with the most advanced fragmentation, as in Western Europe and Southeast Asia (Chisholm 1979: ch. 3; Farmer 1960; Vander Meer 1975). As land becomes scarce, it becomes intensively used and valuable, and people demand permanent tenure. In some African communities increasing land pressure caused by population growth has led farmers to adopt intensive agriculture and has stimulated interest in private ownership. Private tenure has allowed land to become fragmented (King 1977: 345–51; Udo 1965).

The less land people have, the more efficiently they must use it, even if this results in less-efficient use of labor. Farmers maximize sustainable returns to land (value, productivity of land) and accept lower returns to labor (value, productivity of labor) by investing in increased land productivity (Netting 1974a). They stabilize production by planting fruit trees, vineyards, and other permanent labor-intensive crops; and boost yields by applying fertilizer, irrigation, careful horticultural techniques, and by stall-feeding livestock. Unlike swidden agriculturalists, who may operate only one or two fields

at a time (see Cancian 1972; A. W. Johnson 1971), land-hungry intensive cultivators are willing to use many parcels, even small, isolated
ones. The time they spend walking to several fields is rewarded by
access to scarce available land, so they come to regard land fragmentation as a nonissue, if not beneficial.

There are three major advantages to land fragmentation:

Use of several ecozones: in a place with important environmental
variation (e.g., a mountain village) fields in different zones allow
farmers to raise more kinds of crops (Netting 1972, 1981:10–16; Cole
and Wolf 1974; ch. 7; Friedl 1974:56–59; Rhoades and Thompson
1975; Forbes 1976; Weinberg 1972). In Alpine Switzerland, various
crops can be grown at different altitudes within a village's territory.
Swiss villagers need farms in several parcels to have some land in
each local ecozone. Vineyards, gardens, grainfields, meadow, (communal) forest, and alp are all vital to a Swiss mountain household
(Netting 1972, 1981:10–16).

Crop scheduling: scattered fields reduce labor bottlenecks. In a
mountain community crops at lower elevations mature before those
higher up. A farm household that plants the same crop in several
fields, each at a slightly different elevation, can harvest one plot after
another, without hiring outside labor (Netting 1972:134, 1981:18–
21; see also Forbes 1976:246; Cole and Wolf 1974:127–36; Friedl
1974:246; Galt 1979; Fenoaltea 1976).

Risk reduction: land fragmentation also helps avoid risk. Even in
a relatively homogeneous natural environment, such as a large plain,
rain may fall on one field and miss another two kilometers away
(Blaikie 1971). The destructive forces of hail, insect pests, plant disease, tidal waves, and marauding armies may also strike one area and
leave others alone (Bloch 1966:55; McCloskey 1975, 1976; Hyodo
1956). Some fields produce well in some years, while others do well
in other years (Heston and Kumar 1983; Ilbery 1984:164; Carlyle
1983). For example, in one Greek village wet fields produced well in
dry years, while dry fields did well in wet years (Forbes 1976). The
high annual fluctuation in yields of each field is leveled out by having several fields (Forbes 1976; Galt 1979; Netting 1972, 1981: ch. 2;
Friedl 1974:56–59).

There are, undeniably, some drawbacks to land fragmentation, and
whether it is beneficial or not depends on the local economic and
natural environment, and on the land, labor, and capital resources of
each farm. Land fragmentation is most detrimental for farms with

high labor and capital costs. It costs more to move highly paid workers and heavy machines than for low-paid workers to carry their hoes or machetes to another field (O.E.G. Johnson 1970). Larger farmers tend to use more machinery, hire more nonhousehold labor, and have a lower ratio of labor to land, so fragmentation is more of a problem for them than for smaller farmers in the same village. Land fragmentation is less helpful in an area without much ecological variation. In some cases land fragmentation represents an earlier ecological adaptation, which has been rendered maladaptive by rapid economic and technological changes (see Phlipponneau 1975). Yet even in such cases land consolidation is often undertaken thoughtlessly or dogmatically.

Behar (1986:286–303) describes the chaos and emotional crisis of land consolidation in a Spanish village that had had an extremely fragmented land structure. There were 148 holdings in 3,681 parcels (a total of 459.52 hectares and an average parcel size of 0.123 hectares) that were consolidated into 154 holdings in 497 parcels (a total of 422.35 hectares and an average parcel size of 0.850 hectares). So while land consolidation was arguably justified in this case (it allowed combines to harvest the rye), it brought ecological damage and dispossessed the poor and the unemployed. Many villagers lost their garden plots because they could not make the engineers understand the logic of having a little irrigated land close to home. When asked by villagers about the smallest holders who were losing all their land, and where they would raise their vegetables, one engineer replied, "Let them go buy them in the plaza in León." The protective hedgerows and trees were torn down and roads were bulldozed to every field, eliminating 37.17 hectares of farmland. National unemployment reached 20 percent, yet young people who had migrated to the cities could not go back to the villages because land consolidation and mechanization had made the family system of working the land almost obsolete.

FARM ECOLOGY IN NORTHWEST PORTUGAL

There is little environmental contrast in the rolling hill country of the Minho, and few advantages of land fragmentation. Risk management, crop scheduling, and the use of multiple ecozones are only slightly improved by having scattered fields. Recent technical changes include the widespread adoption of tractors and other machinery (Chapter 3 of this book and Bentley 1987b), further limiting any usefulness land fragmentation once may have had. Land fragmentation is part of a complex pattern of kinship and inheritance behavior that allows large farms to be transferred to succeeding generations without disenfranchising all potential heirs. Fragmented farms are easily divided and recombined.

Villagers in Pedralva recognize only three major land use types: forest and brushland (monte);[3] cropland, or field (*campo*); and garden (quintal). Each farm has all three kinds of land use types, and many parcels have both forest and field, or both field and garden. Even a one-parcel farm may have all three types. Three-fourths of the parish is forested, generally on dry, thin, rocky land far from the farmsteads. Flatter land with thicker soil tends to be planted in field or garden crops. In Chapter 5 we will discuss how fields may be reforested, and how forests may be cleared in response to economic and technical change.

Pedralva's farmers distinguish only two major soil types: heavy soil (*terra pesada*) and light soil (*terra leve*). Heavy soil is deeper, moister, and found in lower elevations (330–90 m above sea level in Pedralva). Repeated manuring makes it high in humus (Stanislawski 1959:48). Light soil is thinner and dryer because it is higher on the hill slopes (390–420 m above sea level). Light soil tends to be closer to the forest edge and has more parent rock material (decomposed granite) and less humus. Locals regard heavy soil as better for potatoes, while rye grows better in light soil, but these crops can be grown in either kind of earth if there is no alternative. Maize, grapes, and grass thrive in either soil. Local categories of land use and soil types reveal the subtle ecological differences of a relatively homogeneous environment.

If land fragmentation in Pedralva does not allow farmers to exploit many ecological zones, it does permit some crop scheduling. Because the light soil dries out sooner after the heavy winter rains, it can be planted about a month earlier than the heavy soil. Yet this advantage is slight, because farmers have other ways to regulate planting and harvesting times. Each field has about a month's leeway for planting. Farmers can plant maize varieties that mature in eighty-five days or in one hundred days. They can harvest potatoes two weeks early, or when they are completely ripe.

Land fragmentation helps some risk management for Pedralva's farmers. In dry years potatoes usually grow better in heavy soil than in light soil, but maize, grapes, and grass are not greatly affected by the parish's microclimatic differences, especially because most land is well irrigated. Potato diseases are managed by planting a potato patch in the corner of several maize fields. Pathogens may strike one potato plot but the rest usually escape damage.

Rarely, a few of the largest farmers complain about traveling to many different fields, and about the work of trimming all of the field edges of their many small fields, but even the larger farmers often divide their fields further by planting several crops in one field, which they would not do if land fragmentation were a great drawback. In the winter a farm's largest field may be divided into patches of rye, grass for green fodder, grass for hay, and year-old kale. In the summer the same field may be planted in potatoes and last year's kale, and two varieties of maize. Although the rectangular fields are larger in the village O'Neill (1987a) studied in Trás-os-Montes, and many landholdings totaled up to forty or fifty hectares, walking time was not a problem or a complaint there either (O'Neill, personal communication).

Thus land fragmentation is necessary for operating a small household farm in many parts of the world, but this is probably not the case in Pedralva. This would suggest that inheritance divisions may be the cause of the parish's fragmented land structure. Our discussion of local variation in farm size, inheritance pattern, and degree of fragmentation provides a counterexample to the prevailing notion that smaller farms are more fragmented and that inheritance is the wedge that splits land.

INHERITANCE
Inheritance Customs

Contemporary northwest Portuguese inheritance patterns evolved from a complex historical interaction of two distinct inheritance traditions. Partible inheritance with no sex bias was formalized in the Visigothic code.[4] Male primogeniture originated in Roman practice and was common for the estates of the nobility, including the rural holdings of petty nobles (Silva 1983:15–23).

Ideally, two-thirds of the estate is divided equally among the children, while one-third may be willed to anyone. In practice, one of the children usually inherits the *terço* (the third, i.e., one-third of the estate).[5] Each child receives an equal portion of the rest of the estate, so the main heir receives one-third plus an equal share of the other two-thirds. (Childless people can leave land to a sibling, a sibling's offspring, or a spouse, by testament.)

Farm families encourage one child to *casar-se para a casa* (marry into the house), that is, bring his or her spouse in to form a stem household (Willems 1962). Older siblings who are not willing to work on the farm under their father's leadership generally do not marry into the house, although doing so guarantees later inheritance of the third. The heir who has married into the house inherits the house itself, and usually the farm buildings too.[6] Other offspring may stay in the household if they do not marry. Landless Minhoto households are more likely to be nuclear, rather than stem family (see Brettell 1986:161) because without land to work and inherit there is less inducement for older and younger couples to live together.

After the young couple marries, but before the inheritance division, they gradually assume management of the estate. This time is used to save money and sometimes to buy land in their own name. When the inheritance division is made, the coheirs[7] may receive cropland, forest, money, or some of each. Large farm households strongly prefer to pay off the coheirs in cash. Silva (1983) documents the impartible inheritance of large farms in the nearby Douro Litoral where coheirs are paid off largely in cash. Some large farms in Douro Litoral are known formally as impartibly inherited farms (*casas doadas*, or donated houses), in which the farm is transferred at marriage instead of at the death of the donating owner. In the Minho and in Trás-os-Montes transfer is at death rather than at marriage (O'Neill 1983, 1987a).

Case Histories

The following case studies illustrate the inheritance practices outlined above and show how land fragmentation is generally avoided. Sometimes the estate is divided, especially when the farm is small,[8] but large farms are usually inherited whole in spite of an ideology of partible inheritance (see Cole and Wolf 1974). "Of course, we do well to steer clear of facile dichotomies between partible and impartible inheritance" (Behar 1986:54). The first five cases describe large farms. Unlike most households in Pedralva, these five have enough land to support the male heads of household in full-time farming. According to ideal and practice in the Minho, even secondary heirs

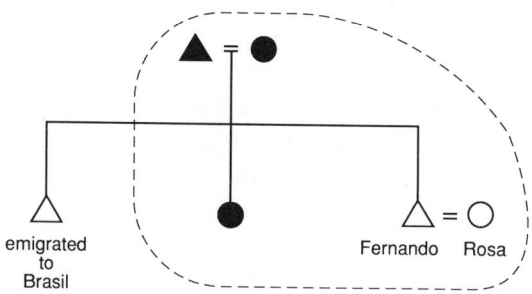

FIGURE 4.1. Case One Kinship Diagram

receive something. As Bourdieu (1976) says when describing a similar inheritance system for France, a small number of implicit principles have spawned an infinite number of practices, which follow their own pattern though they are not based on obedience to any formal rules.

Case one. When Fernando Rodrigues[9] married Rosa Vieira in 1967, his older brother had already emigrated to Brazil (see Figure 4.1). Rosa moved into Fernando's house, where he lived with his parents and invalid sister. Fernando's father died a few years later, and the farm (about 2.3 hectares of fields and 8.6 hectares of forest) was divided among the three children. Fernando inherited the third plus one-third of the rest of the farm. Formally, Fernando received 56 percent of the estate, while each of the other siblings received 22 percent, but in practice the farm was never divided. Fernando's sickly sister stayed in the household, and her share of the land was worked jointly by the household members. The brother in Brazil accepted cash for his share, but since Fernando did not have the money, agreed to wait for it until Fernando could raise the funds. In the meantime the brother loaned his land to Fernando at no cost. A few years later the sister died, leaving her share of the farm to Fernando. In 1984, the year his mother died, Fernando paid off his brother. Fernando solidified control of his parents' farm at age thirty-nine, after being married seventeen years, relatively quickly by local standards.

Case two. In 1984 Fernando's parents-in-law were both still economically active, and Rosa had not yet inherited anything (see Figure 4.2). Rosa's sister was married to another large farmer in the parish, while her brother João had married into the house and lived with his parents, his wife, their two children,[10] and his father's celibate brother. The household farm has about 5.6 hectares of fields and 14.3 hectares of forest. João will probably eventually inherit all of his uncle's share of the farm, one-third of his parents' portion, and one-third of the rest of the farm, giving him the majority of the farm, yet at age thirty-nine João has still received no formal title to the land. He and his wife work the farm with the three elderly household members and one day will inherit the bulk of it.

As this case demonstrates, although elderly people may divide their estate (*fazer as partilhas*) in their own lifetime, they often choose not to. The children who marry out of the house often receive neither land nor cash until they are grandparents themselves, while the heir who has married into the house gradually assumes control of his or her parents' farm without having title to it. A part of the farm profits may be saved each year so that the main heir can pay off the coheirs when the inheritance division is made.

Case three. In the early twentieth century, Maria Fernandes was invited to come live with her five celibate uncles and aunts. The aging siblings wanted an heir for the farm and agreed to will the whole estate to the young girl in exchange for old-age care. She eventually married and had two daughters, Carmen and Glória (see Figure 4.3). In 1964 Carmen (the older) married the only child of a large farmer and moved into his household. In 1967 Glória married the son of a large farmer from a nearby parish, who eventually inherited cash. After her husband died and both of her daughters married,

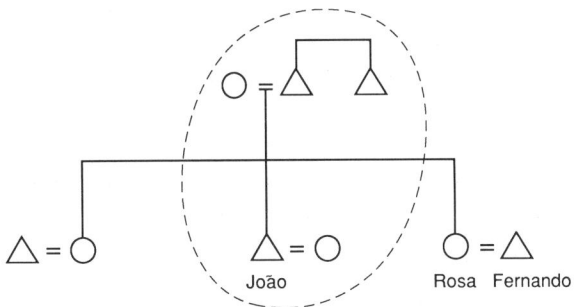

FIGURE 4.2. Case Two Kinship Diagram

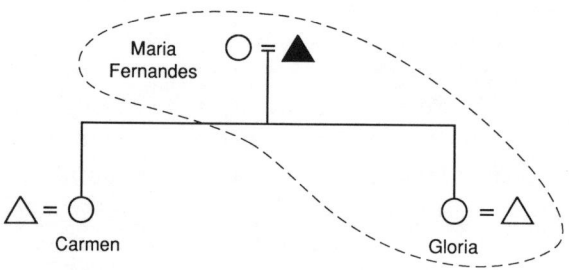

FIGURE 4.3. Case Three Kinship Diagram

Maria divided the estate, giving Glória the whole farm in Pedralva and Carmen another farm in a neighboring parish, let to sharecroppers.[11] In 1984 Glória's farm was the largest in the parish, with about 7.7 hectares of forest and over eight hectares of fields, 2.3 of which had been cleared from the forest. In this case, because there were only two heirs and a large amount of land to divide, each was able to inherit a large farm. When a family has three or more heirs, providing land for all of them is more difficult.

Case four. The Silva family had two sons and a daughter in the 1930s (see Figure 4.4) The daughter was sent to school and became a teacher. In 1958 the older son, Francisco, emigrated to France, returning every year for an August visit.[12] In 1961 Francisco married the niece of his father's brother's wife, who lived with that same uncle. The same year, the youngest, António, married the daughter of a local farmer, and she moved into his parents' home. His wife eventually inherited four small fields—a total of 0.5 hectares—that were worked along with António's father's land. After marrying, Francisco spent his August vacations with his uncle, who was married but childless. He continued working in France, and his wife and uncle continued working the farm. In 1963 António emigrated to France, and his wife stayed with his parents. In 1968 António and Francsico's mother died, and in 1969 António's wife and two children joined him in France, leaving his father alone with the farm. She returned to Pedralva in 1971, and António came home the following year.

Francisco did not return from France until 1977. After seventeen years away from home he assumed responsibility for managing his uncle's farm. His elderly uncle sold him some of the land and gave him title to the rest of the farm (3.7 hectares of fields and 6.8 hectares of forest and pasture) in exchange for old-age care. In 1983 Francisco and António's father died, leaving a one-parcel farm of over four hectares of fields and thirteen hectares of forest and pasture. António inherited the third plus a third of the remaining two-thirds (56%), while Francisco and their sister were each to receive a third of the two-thirds (22%). In 1984 António used what was left of the savings from his French construction job and sold a cow to pay off his siblings in cash, even though Francisco would have preferred land. Because he was fifty years old when his father died, António already had a married son (living in Canada) and four other children at home when he finally inherited his father's farm. Within a few years they would begin to marry and make further demands on the limited farmland.

Case five. Case four shows the pains that people take to avoid dividing large farms, even when there are three heirs. Smaller farms may also be inherited intact, but when there is less to inherit the inheritance may disenfranchise some heirs. The Gonçalves children, Teresa, Madalena, José, and Clotilde, were born in 1928, 1933, 1934, and 1936 in that order (see Figure 4.5). Their parents owned a large house and a single field of 0.93 hectares. In 1957, at age twenty-three, José emigrated to France. In 1961 he married a girl from a neighboring parish and she joined him in France. In 1970 Madalena married a local man who had inherited a house and a small (0.14 hectares) garden plot. Madalena's husband worked in a warehouse in France from 1964 to 1978, although she did not join him in France. The year

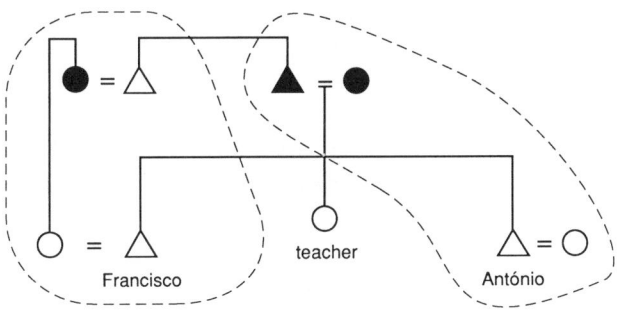

FIGURE 4.4. Case Four Kinship Diagram

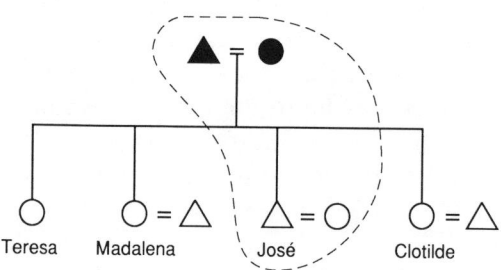

FIGURE 4.5. Case Five Kinship Diagram

he came home they used his saved wages to buy another small house with a 0.182 hectare plot.

José's wife returned from France in 1974, the year his mother died. His wife worked the family farm and kept house for his father until he died in 1979. In 1981 José returned to Portugal, after twenty-four years in France. He paid his sisters in cash for their share of his parents' estate. His wife's parents, who were still alive in 1984, owned a field of 0.34 hectares, which they loaned to José's household. José also farms a plot of 0.11 hectares, which Clotilde bought from a wealthy farmer with money that she and her husband earned in France. She still lives in France with her husband.

The oldest daughter, Teresa, never married but did not remain in her brother's house, as the celibate siblings of larger farmers often do, suggesting that celibates live with their siblings when they are infirm, or when they are needed for their labor and their share of the land, and not for sentimental reasons. Teresa bought a tiny house with no garden and worked as an agricultural day laborer. Her brother allows her to use a few square meters of land in his field as a garden.

Thus even this small, one-piece farm was not divided, even though the coheirs received no land. One of the three coheirs arranged an economically acceptable marriage, another emigrated, while another lives alone in poverty. The one-piece farm was not divided by inheritance but fragmented through growth, as José borrowed a field from his parents-in-law and another from his sister.

All but one of the above five cases describe the inheritance of farms with over two hectares of cropland, large by local standards. Pedralvans call these and similar estates *as grandes casas da lavoura* (the big farmhouses). In each case some strategy was used to prevent dividing the farm. In two cases an in-marrying wife brought land into the household, increasing the degree of land fragmentation as well as the size of the holding. All of the farms described in these case histories were transferred by inheritance, and in one case, by borrowing land. Smaller landholdings are more likely to be acquired through a mixed strategy of inheritance, purchase, and rental. The following case history, like the previous one, presents an example of increased fragmentation through growth of the farm, rather than through division.

Case six. Pedro Vaz was the son of a landless family from a neighboring parish. In 1957 he married a Pedralva woman. Neither of them inherited any land (see Figure 4.6). In 1961 Pedro emigrated to France, where he worked for the next ten years. In 1963 he bought a house and a 0.008 hectare (eighty sq m) garden and a 0.4 hectare field. Over the next few years he bought 2.5 hectares of forestland, and in 1969 he rebuilt the old house. In 1974, three years after returning from France, he rented a 0.13 hectare field from a local man who had emigrated to France. All of his household's cropland—in three separate plots—is slightly larger than half a hectare and was obtained mainly through purchase, using emigrant savings.

While emigration has allowed some people to get small farms (bought from absentee owners, emigrants, or other villagers) their holdings are rarely, if ever, as large as inherited farms. Their smallholdings allow them to supplement their savings and pensions with farm produce and give them some of the landowner's prestige. For many returned emigrants, working on their own land is the culmina-

FIGURE 4.6. Case Six Kinship Diagram

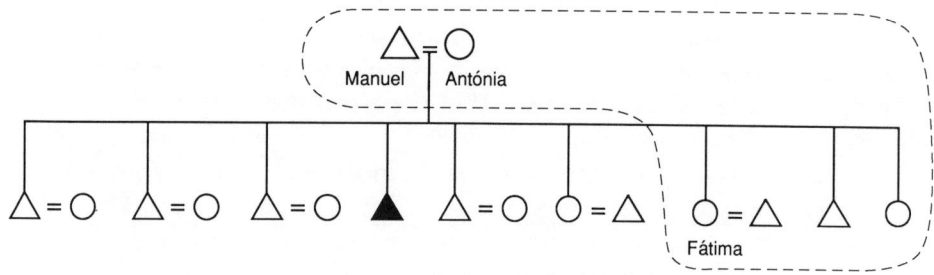

FIGURE 4.7. Case Seven Kinship Diagram

tion of a lifelong dream. For nonemigrants who inherit little or nothing, renting dispersed parcels is an economic necessity, as the following case suggests.

Case seven. Manuel and Antónia were married in 1947, the year Manuel's father died (see Figure 4.7). Antónia came from a neighboring parish and brought no land with her. Manuel inherited his parent's house and and a small field, 0.108 hectares, in 1971, when his mother died. Manuel never emigrated, working locally as a day laborer. Since the early 1970s the household has rented three small fields, totaling 0.318 hectares. In 1984 they rented another field of 0.21 hectares, giving them a total of 0.636 hectares, which Antónia works with her married daughter Fátima. Four older sons and an older daughter have married and moved into their spouses' households. An eighteen-year-old (single) son and a twelve-year-old daughter remain. The small farm, in five fields, was obtained more by renting than by inheriting. The five fields are not the end result of numerous inheritance divisions, but of attempts to add land to a tiny holding by renting several individual fields.

The case studies show that, contrary to received wisdom, land is not fragmented through inheritance. As O'Neill (1987a:164, 318) observed in Trás-os-Montes, subdividing plots at inheritance is very rare; more common is for different heirs to obtain whole, separate parcels, or for heirs remaining on-farm to borrow, rent, or purchase the fields of their nonresident coheirs. "The entire edifice of private property would surely have crumbled and total impoverishment been the lot of every heir . . . if, as is commonly thought, the fragmentation of property through partible inheritance were as unceasing and relentless as the unfolding of time and the generations" (Behar 1986: 34). In the case Behar (1986: ch. 4) describes for León, Spain, villagers generally avoid dividing fields even though farms are divided.

Bourdieu (1976) suggests that marriage rules are like the rules of a game, not like the rules of law. People are dealt good hands or bad and then play them well or badly. In northwest Portugal, the game rules of inheritance call for a modified version of partible inheritance, yet, as in the Romansch-speaking village in the Italian Tyrol (Cole and Wolf 1974), successful primary heirs play the game to inherit a whole farm. Statistical analysis supports the case studies, suggesting that larger farms are more fragmented than smaller farms, and that larger farms tend to be acquired through inheritance. It also shows that, contrary to the common assumption, fragmentation does not harm farm production.

Quantitative Analysis

In Table 4.1 the land acquisition strategies of 158[13] parish landholdings (household farms and gardens) are broken down. Most of the land is farmed by a few large owners; 59 percent of the land in Table 4.1 is owned or rented by twenty-four (15%) of the households. Farms of all sizes rent, buy, and inherit land, but the larger the farm, the greater its inherited portion. Farms over four hectares inherit an average of 5.2 hectares, or 91 percent of the farm. Except for the tiniest holdings, each farm class rents in[14] an average of about one-third of a hectare. The amount of land bought increases steadily with farm size, except for the largest class of farmers, who buy very little land. These largest farmers face the greatest labor constraint and have invested more heavily in labor-saving machinery than in land; they also need less land since they inherited more than other farmers (Bentley 1987b).

Measures of land fragmentation. In Table 4.2 the relationship between farm size and degree of land fragmentation is laid out. Because of the importance of dairying in the local economy, holdings without cows have been deleted, nearly eliminating the ones below 0.5 hectares from discussion.

King and Burton (1982:476) list six major parameters of land fragmentation: farm size; number of plots; size of plots; distance of plots; size distribution of the plots; and shape of plots. Because no

Table 4.1. Land Acquisition Strategies by Farm Size

Farm Class (in ha)	Number of Farms	Total Amount of Land (in ha)	Mean Farm Size (in ha)	Mean Amount Inherited (in ha)
0–0.4999	84	13.77	0.16	0.05 (31%)
0.5–0.9999	28	20.99	0.75	0.32 (43%)
1–1.9999	22	30.93	1.41	0.54 (38%)
2–3.9999	16	48.69	3.04	1.93 (65%)
4 and above	8	45.70	5.71	5.20 (91%)

NOTE: Percentages may not equal 100 because of incomplete data.

Table 4.2. Degree of Land Fragmentation by Farm Size

Farm Class (in ha)	Number of Farms (in ha)	Mean Plot Size (in ha)	Mean Number of Plots	Mean Januszewski's Index
0–0.4999	5[c]	0.25	1.8	.82
0.5–0.9999	23	0.28	3.6	.62
1–1.9999	21	0.32	5.3	.50
2–3.9999	16	0.58	8.2	.44
4–8.5	8	0.64	11.5	.39

[a]Total distance is defined as the sum of the one-way road distance from the farmstead to each field.

[b]Size-weighted distance is defined as the sum of each one-way road distance (in km) from the farmstead times field size (in ha).

[c]Most of the holdings recorded for this category in Table 4.1 were deleted here because they do not have cattle.

one measure (index) of fragmentation captures more than one or two of the major parameters, fragmentation should be measured in several ways.

I have chosen five different indices of land fragmentation: mean plot size; mean number of plots; Januszewski's index of land fragmentation;[15] total distance;[16] and size-weighted distance.[17] The first two indices—mean plot size and mean number of plots—are each direct measures of one parameter. I have chosen two measures of distance (total distance and size-weighted distance) because distance

Mean Amount Rented (in ha)	Mean Amount Bought (in ha)
0.07 (44%)	0.04 (25%)
0.26 (35%)	0.17 (23%)
0.31 (22%)	0.41 (29%)
0.40 (14%)	0.71 (25%)
0.38 (7%)	0.14 (3%)

Mean Total Distance[a] (in km)	Mean Size-Weighted Distance[b]
0.44	0.09
1.50	0.38
2.64	0.80
5.04	1.70
6.53	2.66

is more subjective: it may be measured in a straight line or along roads, from farmstead to each field or from field to field (Igbozurike 1974; Schmook 1976; Dovring 1965:40–42). Unfortunately, no index measures field shape.

Common opinion holds that smaller farms are more fragmented than larger farms. In Table 4.2 we can see that the larger, mostly inherited farms are more fragmented than the smaller farms, as measured by four of the five indices. The only index in which larger farms appear less fragmented is mean plot size, but the increase in

Table 4.3.   Tractor Ownership by Farm Size

Farm Size (in ha)	Number of Households	Number of Tractors	Average Number of Tractors per Household
0	95	1	—
0.0001–0.4999	84	1	—
0.5–0.9999	28	2	0.07
1–1.9999	22	8	0.36
2–3.9999	16	11	0.68
4 and above	8	11	1.38

plot size is not proportionately as great as the increase in farm size, so distance and number of plots increase as farm size increases. The largest farms are the most fragmented,[18] having the most plots, the lowest Januszewski's index, the greatest total distance, and the greatest size-weighted distance.

Although the largest, inherited farms are the most fragmented, inheritance does not cause land fragmentation. As the case histories indicate, most large farms are inherited intact, so that the actual division of fields or farms rarely takes place. Inheritance transfers large, already fragmented farms, rather than fragmenting further, and farmers rarely buy or trade land to consolidate their holdings. Farms become more fragmented as they enlarge their holdings by renting and buying separate fields, and as in-marrying spouses bring in more fields. A whole farm sold in 1969 was bought by three landless (or nearly landless) households, which each got about a hectare of land, Large farms are rarely as severely divided by inheritance as this farm was by sale.

Farm production. The larger farms are not only more fragmented and more likely to be inherited whole than the smaller farms, they are more likely to have a tractor and an adult male household member who works full time as a farmer. The larger farms have more machinery, more capital, and less labor per unit of land than smaller farms. In Table 4.3 tractor ownership is matched to size of holding. Several nonfarmers own tractors (for custom work), so all parish households were included in the sample.

Small farmers feed their cattle grass and maize tassels and leaves, keeping the grain for the family's bread. Farm families with more land are not in such direct ecological competition with their cattle because they can raise all the grain for bread on a portion of the land, dedicating much of the rest of the farm to maize silage for dairy cattle. With more land, more capital, and less pressing competition between people and animals over basic food, one would expect larger farms to raise more cattle per hectare and to produce more milk per dairy cow than the smaller farms. Carvalho, Barros, and Rocha (1982:23–24, 37) decry land fragmentation as a major limiting factor in northwest Portuguese milk production, typical of the prevailing dogmatic attitude about land fragmentation. And in Table 4.4 we see that the larger farms indeed produce more milk and raise more cows than small farms. However, cow-carrying capacity, milk production per animal, and mean milk production per square meter of farmland is fairly similar for each class of farm. Large and small farmers achieve the same yield per unit of land. Because of their intensive use of household labor, small farms produce a comparable amount of milk per unit of land as large farms, in spite of having to devote a larger proportion of the farm to household food production.

Farms between two and four hectares have the highest milk yield per animal, perhaps because they have the most mature herds. Large farmers went into dairy production before smaller farmers, generally in 1978 and 1979, so by 1983 and 1984 their herds had relatively few heifers and calves (Bentley 1987b). The largest farms are still expanding their herd size, so they have less-mature herds than farms of the two to four–hectare group.[19]

Possible error. While there are a few possible sources of error in the data, they do not shed serious doubt on the results. One source of error is home consumption of milk, not measured in this sample. About half of the dairy-producing households take home one-half to one liter of milk daily. This "subsistence off-take" may total over a hundred liters a year, still a fairly small amount. A second possible source of error is that dairy production is new (since 1978) in Pedralva, and farmers are still adjusting to it. When dairying was introduced commercially, over twenty households were already rearing one or two milk cows for home production, but no one kept the maximum number of dairy cattle. Because dairy cows eat more than

Table 4.4. *Dairy Output by Farm Size*

Farm Class (in ha)	Number of Farms[a]	Total Milk Yield[b]	Mean Milk Yield[c]	Mean Herd Size[d]	Yield per Cow[e]	Yield per Square Meter[f]	Cow-Carrying Capacity[g]
0–0.4999	3	4,700	1,567	1[h]	1,567	.50	2.85[h]
0.5–0.9999	22	66,700	3,032	1.6	1,997	.39	2.06
1–1.9999	21	89,600	4,267	3.2	1,506	.31	2.25
2–3.9999	16	148,800	9,300	5.8	2,078	.31	1.89
4–8.5	8	176,400	22,050	13.4	1,579	.34	2.15

[a] Number of farms is limited to those that sell milk. Three farms own cattle but do not sell milk (two in the smallest class, and one in the next largest class).

[b] This is the total liters of milk marketed in one year for all farms in this class.

[c] This is the mean number of liters marketed in one year per farm.

[d] Mean herd size includes immature animals and work cows.

[e] Yield per cow is annual and includes immature animals.

[f] This figure gives the average liters of milk marketed in one year per square meter of cropland.

[g] This is the average number of cows, including immature animals and work cows, per hectare of cropland.

[h] These figures are based on five farms in this category.

work cows, no one knew exactly how many dairy animals his farm could sustain. Larger farmers were especially uncertain about how many cows they could raise because, before the advent of tractors in the 1960s (Bentley 1987b), large farms had never had more than eight or nine work cows, enough for two complete plow teams of four cows each. By 1978 few were raising even that many. They could have kept more cows if they had cared to, but they had only as many as they could use for spring plowing. When commercial dairy production was instituted the larger farmers began raising more cows than they had ever raised before, cautiously adding a few animals every year. Some of the largest farms are still expanding their herd size and have many young animals. When their herds mature they will have greater milk production. Smaller farms reached their full dairy potential soon after dairying was introduced. That animal productivity of the eight largest farms is much less than that of the next smallest farm class, reflecting the immaturity of larger farmers' herds, is clear from Table 4.4.

I used milk yield to represent total farm productivity because heavy government subsidies have made milk more profitable (Finan 1987), and farmers tend to produce as much milk as possible after satisfying household demands for bread. Milk production is therefore a useful indication of overall farm production. I was fortunate enough to get complete dairy production figures from the local dairy cooperative and thus have better data for milk than for other commodities.

LAND FRAGMENTATION AND FARM PRODUCTION

If farm fragmentation is detrimental, it should lower milk output. Increased distances would waste farmers' time, so that many hours would be spent walking instead of in more-productive activities (see Karouzis 1971). Farmers would abandon or disintensify far-flung plots. The many small plots of the highly fragmented farm would present a labor constraint at plowing time, so that some land could not be planted in maize and would be left in summer pasture.

As indicated by Table 4.4, large farmers raise more cows than small farmers do, but only in proportion to their larger farm size. There is a high correlation between farm size and herd size (r equals .80; significant at the .0001 level). When scale-neutral measures of production are compared to indices of fragmentation, no correlations

Table 4.5. Regression Analysis of Scale-Neutral Milk Production Variables with Indices of Fragmentation

	A. Simple Regression	
	Milk Yield per Cow with	Correlation Coefficient
Variables	Number of plots	.03160
	Januszewski's index	.02306
	Total distance	− .03460
	Size-weighted distance	− .06720
	Milk Yield per Square Meter of Land with	Correlation Coefficient
Variables	Number of plots	.05184
	Januszewski's index	.08602
	Total distance	.03320
	Size-weighted distance	− .08279

	B. Multiple Regression		
Dependent Variable	Independent Variables	Significance	Overall *F*
Milk yield per cow	Farm size	.36179	.873
	Number of plots		
	Januszewski's index		
	Total distance		
	Size-weighted distance		
Milk yield per square meter	Farm size	1.24962	.297
	Number of plots		
	Januszewski's index		
	Total distance		
	Size-weighted distance		

are found. The relationship between land productivity and land fragmentation is random.

A regression analysis was devised for seventy farms with complete land and milk data, using scale-neutral measures of farm production (annual amount of milk produced per dairy cow, and annual amount of milk produced per square meter of farmland). In Table 4.5 we see a complete lack of correlation between scale-neutral measures of land productivity and production per cow with farm size and four indices of land fragmentation. The overall *F* for milk yield per cow with the indices of fragmentation is .32506 (significance of .896); the overall *F* for milk yield per square meter of cropland is .99691 (significance of .427)—that is, not at all significant for either score.[20]

The critics of land fragmentation assume that fragmentation is so detrimental to farming that it lowers land productivity. The farmers supposedly waste so much time moving from plot to plot that they do not have enough time properly to cultivate any of their land (O'Flanagan 1980; Karouzis 1971; Burton and King 1982; Jacoby 1971; Meliczek 1973; Naylon 1959; King and Burton 1982; Lambert 1963). Discussing land fragmentation in the Minho, the Portuguese anthropologist Pina-Cabral (1986:22) wrote, "Such great subdivision of the terrain is certainly uneconomic, not only because of the land taken up by paths and hedges and because of the length of time required to reach different plots of the same farm, but also because it prevents the use of complex agricultural machinery and a consequent adoption of modern agricultural methods." Although he goes on to give a more balanced perspective of the issue, Pina-Cabral is convinced that land fragmentation is dysfunctional. My study suggests that fragmentation does not decrease yields. The greatest problem with criticisms of land fragmentation is that the critics fail to measure land or labor productivity. Land fragmentation is assumed to be so detrimental that empirical analysis is deemed unnecessary.

Although critics claim that land fragmentation is Europe's greatest detriment to agriculture, land fragmentation neither hinders nor increases agricultural production in Pedralva. In the Minho land fragmentation is agronomically neutral, a nonissue. There is no common folk term either for fragmentation or for consolidation. Large, inherited farms are the most fragmented, but inheritance histories suggest that neither fields nor farms are commonly divided through inheritance. A farm of ten fields or a farm of one field is usually inherited by one heir who buys out coheirs with cash, not with land. Smaller farms may become more fragmented with time as farmers take advantage of the fragmented land structure to acquire land gradually, according to their ability to save money and exploit their own labor (see Brettell 1982:50, 77). Small fields are the small change of the land system, and they make it more flexible and resilient (Robert Netting, personal communication). However, I am not suggesting that land fragmentation exists to provide small farmers with small increments of land or to allow a married couple to pool their separately acquired land. Such functions are compatible with land fragmentation but are not its reasons for being.

As the Nigerian geographer Igbozurike (1974) says, land fragmentation is an overrated phenomenon. It is a correlate of intensive agriculture and individual land tenure. It plays a minor role in adjusting farm size through the operation of the land market and by allowing a married couple to pool their separately inherited fields. Fragmentation allows a finer adjustment of land to individual requirements and stages in the household development cycle than would be possible without it, and it probably allows a larger number of people to be sustained from the land. Nonetheless, land fragmentation in Pedralva is not an essential part of the ecological adaptation, as it is in other, more complex natural environments (Netting 1972, 1981; Galt 1979; Forbes 1976; Friedl 1974). Larger farms get most of their land through inheritance and are more fragmented than smaller farms, but produce no more milk and cows per square meter than smaller farms.

Portuguese agronomists complain that they can do nothing to improve Portuguese agriculture until the small farms have been consolidated (Carvalho, Barros, and Rocha 1982; Portela 1981). Yet my analysis suggests that land fragmentation is not an agricultural problem in northwest Portugal because of the relatively homogeneous natural environment and low labor costs. The underlying, historical cause of land fragmentation in the Minho is probably the hilly nature of the terrain (Pina-Cabral 1986:6, 8). The terrain limits twentieth-century field clearing to fairly small parcels (see Chapter 5). Land fragmentation neither helps nor hinders the Minhoto farmer. The policy implication is that consolidating northwest Portugal's farms would be expensive and useless.

The Center of Pedralva as seen from a nearby mountain. Gorse and other moorland plants are in the foreground. Forest surrounds the farmland and large villages of the parish center. The bold lines between fields are grape ramadas. (All photographs are by the author.)

Loading Brush. A woman tosses a basket of ground litter to workers on top of the load, who catch it with rakes.

Composting Brush. Making a large pile of brush, covered like an iced layer cake with manure, provides crucial nutrients to summer maize planted in fields just after winter grass has been harvested.

Weeding. A farmer walks briskly to match the horse's pace, holding the cultivator up to cut weeds and break the soil between young maize plants. Horses are valued for their speed and because their hooves, unlike a tractor's wheels, do not pack down moist soil.

Stacking Maize Stalks. After removing the ears, a farm woman lifts bundles of stalks to her husband who ties them to a post. Work on ladders is one of the few field tasks generally reserved for men.

Ensiling Maize. Workers snatch a brief rest while the tractor driver raises the hydraulic lift on the wagon. Maize is tightly packed into the wagons, which means that workers must dig the chopped maize out of the wagon with rakes before spreading it out around the silo.

Threshing Rye. Large crews gather for the hard work of rye threshing. The farmer (*on top of the machine*) helps feed the rye into the thresher as others rush to carry off the straw, chaff, and clean grain. Grapevines surround the threshing floor.

Planting Intercropped Potatoes and Kale. Wilted kale seedlings, lying in two widely spaced rows, will revive and flourish. In the background, piles of manure wait to be spread on the grass before the sod is turned. Posts for a grape arbor stand around the plot, and behind sits a two-storied grain barn.

The Grape Harvest. Grapes, in the form of wine, are an important carbohydrate source in the people's diet. The two large wooden casks on the wagon are dornas, used by most farmers to collect the harvest and sometimes used by poorer farmers to ferment the wine in.

An Old Persian Wheel, or Noria. A farmer has placed a fallen log in the horizontal wheel to show how cattle were harnessed to it to draw up water in a continuous line of buckets. The metal pails are no longer on the vertical wheel.

Loading Hay. The draft power is provided by dairy cows, whose milk yields, despite the fears of agronomists, are not substantially lowered by their innovative role on Portuguese farms. Grape ramadas surround the field.

Intensive Land Use. A sharecropper holds a cow on a tether to keep her out of the haystack (*left*) and the maturing rye (*right*) as she grazes a narrow band of grass. Farmers grow crops both on the ground and "in the air." The centerpost is a granite esteio, a support for a grape ramada.

5. Land Use Changes and the Ecology of Field and Forest

Several Portuguese agronomists I met expressed a concern that emigrants were buying fields and "abandoning" them, that is, not farming them, allowing the parcels to return to forest. Small woodland owners in general are often asumed to be less than professional managers of their lands (J. P. Jackson 1984; Birch 1986; Towell 1982). A main concern is that trees are not "harvested" at the optimum time. "Small woodland tracts by and large remain outside the sphere of economic management" (Towell 1982:192). Some foresters describe small woodland owners as careful managers; Hilliard-Clark and Chesney (1985) discuss black nonindustrial private forest land (NIPF) owners in North Carolina as thoughtful, rational managers of their woodlands. Even though they have no contact with formal forestry programs, the owners rely on their life's experience for important technical skills.

Midway between these perspectives there is an attempt to identify only some of the small woodland owners as good managers. Young and Reichenbach (1987) found that Illinois NIPF owners who intended to harvest timber were more likely to be interested in forest management. Romm, Tuazon, and Washburn (1987:198) concluded that NIPF owners in North Carolina were more likely to invest in forestry if they were young, full-time residents (not absentee owners) with high income: "rational economic choice, while not the only conceivable explanation of NIPF owner behavior, is a useful policy assumption and a fruitful hypothesis for further study."

I seek to demonstrate here that Pedralva's forests are managed according to rational economic motives, and that emigrants are not generally responsible for land "abandonment." It is more important, however, to describe field-forest changes in detail and interpret them according to historical changes and intracommunity variation. There is quite a bit of cropland that has reverted to forest over the years, as well as forest that has been cleared for fields. The kinds of plots cleared or reforested have changed through time with innovations in farm technology. Large farmers and small landowners have different patterns of land use changes.

The average farm has 6.3 hectares of land, in 13.3 parcels, many of which have both field and forest. Since most of the land is in forest the average amount of cropland per farm is only 1.9 hectares, divided into an average of 5.5 different fields. The desire to farm more land leads aggressive young farmers to clear patches of forest, while fields that offer low returns to labor and capital (especially machinery) are allowed to reforest naturally or are planted in trees. Reforested land is not abandoned; it has been given over to a less labor-intensive strategy and reflects a labor-saving decision. Reforested land still yields brush, grazing, firewood, and timber, and retains value as real estate. Land reforestation increases returns to labor (for that plot), while lowering returns to land.

Land use changes are an adjustment to the technical changes discussed in Chapter 3, especially the adoption of dairy cattle and water pumps, and the switch from yellow cows and oxen to tractors. New farm machinery—especially the tractor—is better suited to larger fields, enhancing the attractiveness of the new, larger fields, while making smaller or terraced fields less attractive. Farmers can sell off a stand of timber and use the money to pay a contractor to level the land and bulldoze out boulders and tree stumps.

Cleared forest soils are decomposed granite, thin and leached of most nutrients by heavy winter rains (Stanislawski 1959:141–66). Such lands were too sterile to be productive in the early and mid-twentieth century. Now, however, Minhoto farmers have chemical fertilizers to enrich these soils. In addition, the growth of commercial dairy farming since 1978 has given farmers more natural manure to use as a fertilizer. The owners of larger farms (and generally the owners of more forestland) are now raising more cattle than before 1978.

The new fields are also easier to irrigate because of new water technology: improved wells, pipe, and pumps. Formerly, wells were dug by hand (through several meters of granite). The soft, water-saturated stone at the bottom of the well was lined with masonry, to prevent the sides from collapsing. Once water was reached, crews of men worked around the clock for days to complete the well. Once a well was finished, water was drawn to the top by a Persian wheel turned by oxen. Water could be moved only downhill from the well through ditches, and occasionally through channels cut in stone, or under fields in *canos* (underground ditches, lined on the bottom, sides, and top with flat stones, and covered with soil).

Now wells are dug by two-man teams, outside specialists who use dynamite and jackhammers to dig to the aquifer (about eight meters below the surface), lowering prefabricated well-casing rings of concrete and steel into the well, casing it as they dig. Electric or gasoline-powered pumps quickly move water through plastic tubing to remote or uphill plots of land. When fields were plowed with oxen, and water flowed only by gravity, small terraced fields on the steep sides of canyons were valued for being near streams. These areas are now being reforested as farmers clear larger fields and pump water up to them.

Larger farmers are more active in reforestation and in land clearing than smaller farmers are. There are probably several reasons: larger farmers have more money for irrigation equipment (tanks, wells, tubing) and fertilizer to make new fields productive. Larger farmers face the greatest labor constraints and may be motivated to stop cultivating less-productive fields. Larger farmers are more likely to own plots of forest that are suitable to be cleared and leveled for fields. While differences in the forest management practices of larger and smaller farmers correlate with differences in their supplies of land, labor, and capital, I am not suggesting that smaller farmers are inherently traditional or conservative. A person who owns half a hectare of woodland lives in a much different economic environment than a person who owns twenty.

REFORESTATION

Case history analysis suggests that reforested plots are most often marginal, distant, small, or without dependable irrigation. The fields that were good field sites when animals were the only source of traction, and water could be moved exclusively by gravity, were no longer rational field sites after tractors and water pumps came onto the scene.

In Table 5.1 reforested land is listed chronologically. (See Figure 5.1 for the locations of the plots listed in Tables 5.1–5.4.) The first column in Tables 5.1, 5.2, and 5.3 is a code number for each plot. The second column is a code number for each owner. Some households own more than one plot of reforested land. The third column is the amount of cropland farmed by each household in 1984, in hectares. This figure is listed to give an idea of the relative size of each farm, to suggest that farm size affects forest-field ecology. Most of the cropland holdings listed in Tables 5.1 and 5.2 are well above the average (1.9 hectares) cropland holdings for Pedralva.

The fourth column (omitted in Table 5.1) lists the years that the current (male) farmer was abroad (not counting military service). In a few cases villagers suggested a relationship between land reforestation and emigration: as emigrants lost interest in maintaining their fields they allowed them to revert to native vegetation. However, emigration is not a major cause of land reforestation. Many reforested plots are owned by households with no emigration history.

The fifth column (omitted in Table 5.1) is the date of land reforestation. The term *early* means early twentieth century. Column six lists the size of the reforested plot—an estimate in a few cases—taken from the 1983 land survey, from a recent 1:5,000 map, or measured in the field.

I was interested in the relationship between changing irrigation technology and changes in field sites. Water source in column seven refers to the source of irrigation water when the plot was cultivated. Ponds are the poças discussed in Chapter 2. Dry means that the plot was watered only by rainfall and probably produced only winter rye (which is not irrigated). Column eight lists the distance in meters from each reforested plot to the farmstead.

Pre–Twentieth-Century Reforestation

The plots reforested before living memory (plots one through fourteen) tended to be dry, marginal plots, owned by larger farmers (see Table 5.1). Plots two through twelve were a group of relatively large, contiguous fields, in an area called Poços. They are now densely wooded in pine, eucalyptus, and oak, between a deep wood and a large field system. Older people claim that the area was once farmed, although they do not know when. The place is crisscrossed by roads, walls, and ditches, as in cultivated areas. There is a small aban-

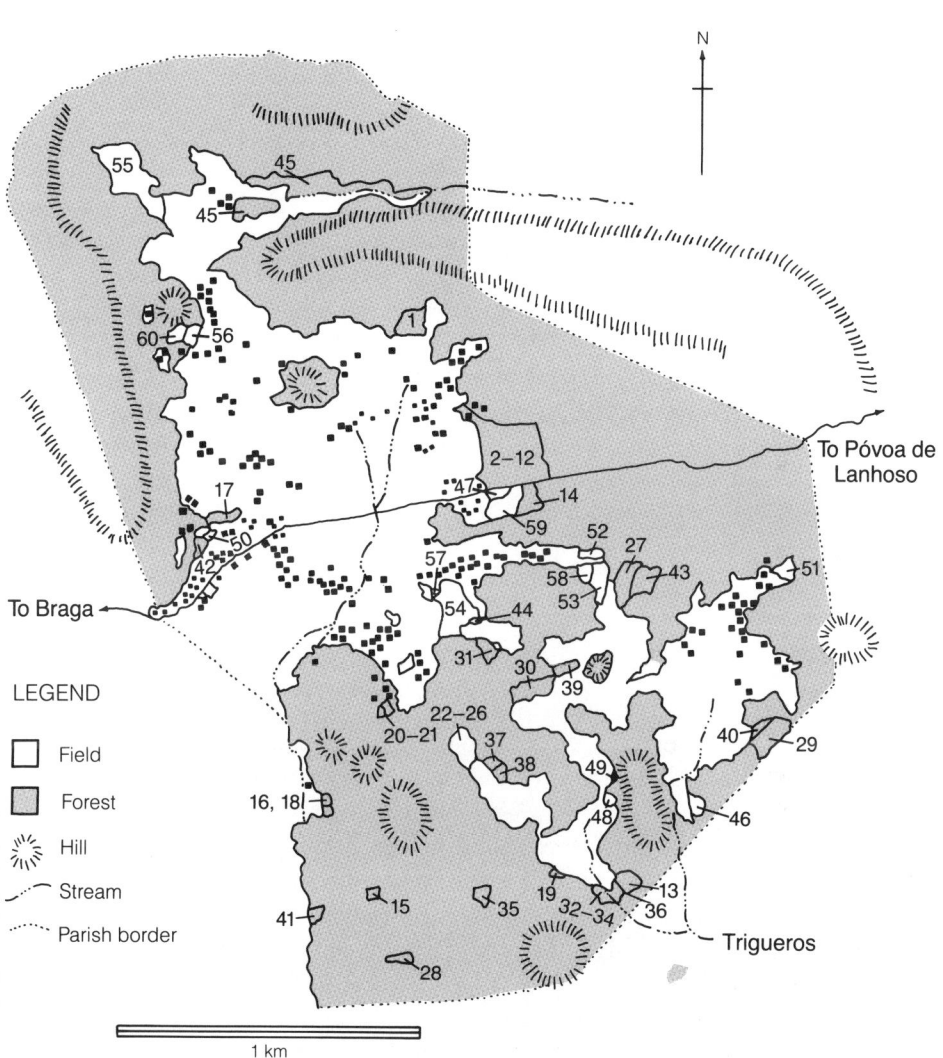

FIGURE 5.1. Locations of Land Use Changes

Table 5.1. Reforested Land before Living Memory

Plot	Owner	Amount of Cropland (in ha)	Size (in sq m)	Water Source	Distance from Farmstead (in m)
1	1	4.9	6,000	well	1,350
2	2	5.6	400	well?	350
3	3	5.4	1,460	well?	350
4	4	??	3,000	well?	??
5	5	2.3	9,600	well?	300
6	6	1.4	9,800	well?	1,100
7	7	3.9	16,000	well?	300
8	8	2.8	7,200	well?	300
9	3	5.4	3,800	well?	350
10	2	5.6	1,400	well?	350
11	9	1.2	7,600	well?	1,550
12	10	4.2	7,500	well?	300
13	11	1.7	4,500	dry	2,350
14	3	5.4	13,470	dry	650
Mean			6,550		740

doned well and an abandoned irrigation pond (both now dry) at the highest edge of the area, which probably once irrigated at least some of it, although there was probably not enough water for all of the fields.

The remaining plots abandoned before memory also had inadequate water supplies. Plots thirteen and fourteen are dry. Plot one was a marginal field bordering a rocky forest area. The field had a well, although probably not a very productive one. Plot one was fairly distant from the farmstead, while plot thirteen was very distant (by local standards), over two kilometers.

We do not know how long this land was cultivated. Because it is marginal land, it may not have been farmed for long. These early episodes of land reforestation demonstrate that some land was reforested before farmers bought tractors and before the massive emigration of the 1960s drained off much of the labor force. The common image of European farmers cutting down forests to make more fields is an oversimplification. These cases suggest that the reverse process also happened, and well before the rapid changes of the midtwentieth century.

Early Twentieth-Century Reforestation

Fields reforested in the early twentieth century (up to and including the 1940s) tended to be smaller parcels, at greater distances, with poor water supplies (see Table 5.2).

Plots fifteen, sixteen, eighteen, and twenty were all dry fields. They were carved out of the forest by poor individuals, worked for a generation, and then reforested. These marginal lands could have supported only rye, and only at great effort. Each of these *tapadas* (enclosures) is associated with an individual's name (e.g., A Tapada de Pedro, Pedro's Enclosure). The person who cleared the forest is still within living memory (or almost). These plots were often entirely surrounded by stony forestland. They represent unsuccessful attempts to eke out a living by farming very marginal lands.

Plot nineteen had adequate water: a day's turn once a week from a small irrigation pond, still used to irrigate a large maize field. Plot nineteen was probably allowed to revert to forest because it was owned by a large farmer who could not afford to travel over two kilometers to tend a small field of nine hundred square meters. In general, larger farmers cannot afford the time to farm land far from their homes. Even in a country as different as Ethiopia, smaller farmers with relatively more labor can afford trips that larger farmers cannot make (R. T. Jackson 1970).

Plots twenty-two through twenty-six were contiguous plots in an area called São Marcos. Villagers claim that São Marcos was reforested gradually between 1944 and 1974, although 1974 aerial photographs show that the area has been covered in forest at least since that year, indicating reforestation well before that time. There is a small irrigation pond (dry in 1984) at the head of the field system, which never could have provided very much water for the São Marcos fields.

The São Marcos area was more intensively cultivated than the four enclosures. Abandoned grape arbors around the edges of the São Marcos fields testify to an earlier investment in labor and money. São Marcos was probably reforested because the fields are small—a few thousand square meters each—and because the one small irrigation pond probably could not water all of them. Although now converted to pine forest, some of this reforested land still bears the narrow furrows of rye cultivation (rather than the broader furrows of maize), suggesting that maize may not have been grown in these fields. Infor-

Table 5.2. Reforested Land, Early Twentieth Century through 1940s

Plot	Owner	Amount of Cropland (in ha)	Years of Emigration	Date of Reforestation
14	3	5.4	0	old
15	12	1.0	1969	early
16	13	0.1	1957–72	early
17	1	4.9	0	early
18	?	??	??	early
19	8	2.8	0	1920s
20	10	4.2	0	1930s
21	?	??	??	??
22	3	5.4	0	1940s
23	14	3.9	0	1940s
24	15	3.4	0	1940s
25	16	retired	0	1940s
26	17	retired	0	1940s
27	9	1.2	0	1940s
28	18	??	??	1940s
29	19	0	permanent	1940s
30	20	0	permanent	1940s
Mean				

NOTE: Plots 22–26 (São Marcos) have a total area of 12,950 square meters. I measured the area of the reforested land, which was still clearly marked by such agricultural features as terraces, abandoned grape arbors, roads, and walls. However, the 1983 land records were too sketchy to allow a precise identification of which areas were owned by which people, although we know the owners who hold reforested land in São Marcos.

mants said that São Marcos generally produced rye. Maize is much more profitable than rye but needs summer irrigation. A field that can grow only rye is not nearly as valuable as one that can grow maize too. Labor inputs are highest for maize, lower for rye, and still lower for fields planted in trees.

Plot twenty-seven was reforested about the same time as São Marcos because it has no irrigation water. It was a large field, near the farmstead and next to another large field owned by the same household.

Plot twenty-eight was a relatively large plot with its own irrigation pond. Although it was entirely surrounded by forest (indicating that

Size (in sq m)	Water Source	Distance from Farmstead (in m)
13,470	dry	650
3,500	dry	2,000
few−100	dry	1,100
2,700	well?	250
few−100	dry	??
900	pond	2,250
2,000	dry	1,500
2,000	dry	??
12,950	pond	1,500
	pond	1,500
	pond	1,100
	pond	800
	pond	1,600
11,460	dry?	750
5,000	pond	??
7,200	dry?	??
8,200	dry	??
3,560		1,300

the soil is thin), it is in a low, wet spot and should have been acceptable cropland. It is on the border of the parish and is owned today by people in Sobreposta (the parish to the west of Pedralva). The owner's relatives in Pedralva are sharecroppers, so it is unlikely that the previous owners were wealthy. Plot twenty-eight was cleared and then reforested within twenty years, suggesting that the land was marginal.

Plots twenty-nine and thirty were both owned by people who emigrated permanently to Brazil in the 1940s. Plot twenty-nine was large, dry, and surrounded by forest. The owners of plot thirty sold their whole farm (to various individuals) before emigrating. Other

Table 5.3. Reforested Land, 1950–1984

Plot	Owner	Amount of Cropland (in ha)	Years of Emigration	Date of Reforestation
31	21	3.8	0	1950s
32	22	0.4	1963–1971	1950s
33	15	3.4	0	1950s
34	23	1.4	1957–1981	1950s
35	24	6.0	0	1950s
36	25	1.5	0	c 1959
37	15	3.4	0	c 1959
38	15	3.4	0	c 1959
39	26	0.7	0	1960s
40	9	1.2	0	1960s
41	27	??	??	1960s
42	21	3.3	0	1960s
43	15	3.4	0	1967
44	28	??	??	1970s
45	29	1.5	0	1980s
46	30	3.2	1964–	1980s
Mean				

fields of the farm were reforested, but plot thirty was dry, although large and centrally located. The plot lies at the border of field and forested areas, suggesting that its soil is marginal.

The plots reforested in the early twentieth century tended to be small, with an inadequate water supply. They were also fairly distant by local standards, in a community where (in 1984) it was unusual for any field to be over two kilometers from the farmstead. These cases suggest that land may be reforested for rational reasons other than environmental characteristics of the plot itself. A farmer may reforest a field to minimize distance, just as farmers grow less-intensive crops on more-distant fields (Chisholm 1979: ch. 3; Blaikie 1971; DeLisle 1982).

Some of the largest plots reforested were owned by permanent emigrants to Brazil. When they emigrated they left land that others either could not cultivate or were not interested in. Between 1930 and 1950, 111 people emigrated from Pedralva (Bentley 1987b:171, Table 9.2). Although two emigrants allowed fields to be reforested, emigration per se is not a major cause of reforestation. Emigrants

Size (in sq m)	Water Source	Distance from Farmstead (in m)
6,900	well	1,600
700	stream	2,300
1,900	stream	2,300
2,200	stream	2,300
8,170	dry?	1,600
2,100	stream	2,350
1,490	well	1,500
2,000	pond	1,500
2,970	dry?	1,650
3,000	dry?	350
2,930	dry?	??
3,600	dry	650
9,000	dry?	1,400
2,800	dry?	??
7,400	stream	650
55,870	well, stream	0
7,060		1,440

wishing to keep a plot of land in their home country could reforest it or could rent it out.

Midtwentieth-Century Reforestation

By the midtwentieth century (1950–1984), the pattern of reforestation had changed dramatically (see Table 5.3). Reforested parcels were no longer always marginal tracts of dry land. Adequately watered parcels were returned to forest when they were small (often terraced) and distant, reflecting the changing technology of midtwentieth-century agriculture. As farmers came to rely on tractors, they found smaller parcels difficult to plow. Machine pumps and new wells increased the supply of water, so farmers placed less value on the narrow but adequately watered plots along the steep stream banks.

Plot thirty-one was fairly large. It now supports a stand of big oak trees and once had some water from a nearby well. It is owned by a large farm household. Distance was probably a factor in disintensifying plot thirty-one, since it is sixteen hundred meters from the farmstead, while the farm's remaining twelve fields are an average of only 475 meters from home. This household has disinvested in land in other ways as well, renting out and selling a few small plots since the 1970s.

Plots thirty-two through thirty-four were contiguous plots in a small area called Tumudas. The Tumudas plots were small terraces on a steep stream bank. Plot thirty-six was a similar tract across the stream from Tumudas, reforested a few years later. All of these plots have a good water supply but are too narrow to work with a tractor. Even though two of the owners are small farmers who emigrated, emigration alone does not explain why they stopped farming these plots. The owner of plot thirty-two emigrated when his wife and father-in-law were operating the land. The owner of plot thirty-four emigrated as a teenager while his father stayed and worked the land. All four of these plots faced the additional disadvantage of being relatively far away, over two kilometers.

Three of the four plots (thirty-two, thirty-four, and thirty-six) were owned by small farmers, those least likely to stop cultivating a plot. Even small owners disintensified land use (by growing trees instead of crops) as labor became scarcer and capital became more abundant. Plot thirty-six is still called A Quinta de Ramos (Ramos's Farm), after the house name[1] of the owners. Dead grapevines sprawl through the tree branches at the edges of the little terraces, evidence of earlier agricultural investment and recent cultivation.

Plot thirty-five is called the Bouça de Pão (Bread Forest) because it once grew grain crops. Owned by one of the parish's largest farm households, it was a marginal plot of fairly dry land, like many of the parcels reforested in the early twentieth century. Although it is now mature pine forest, men in their fifties remember working in it as boys, when it grew rye and some corn. The owners of the Bread Forest once hired much labor and had sharecroppers. They now farm almost entirely with household labor and machinery. Because it was a distant, marginal field, reforesting the Bread Forest was one of this household's first adjustments to rising labor costs beginning in the 1950s.

Plots thirty-seven through forty-three were abandoned for similar reasons as the Bread Forest. Except for plot forty-three they are all fairly small, but large enough to plow with tractors. Plots thirty-seven and thirty-eight had a little water. Plot thirty-seven was watered from a small well (now dry), while plot thirty-eight was irrigated with water borrowed from a farmer who owned a neighboring field. As mentioned in Chapter 2, borrowing water entails return obligations that people sometimes find onerous.

Plots thirty-seven through forty-three were all either in the forest or at the edge of it, that is, on marginal land. Several of these plots were owned by a single household (owner fifteen) who disintensified the use of marginal land as labor costs rose. Disintensification does not necessarily indicate a lack of interest in agriculture. This point is best illustrated by the case of plot forty-three, which was bought as a field from a farmer who had emigrated to Brazil. The new owner immediately planted the field with eucalyptus trees, investing in land while consciously disintensifying its use. Larger farmers are less land-hungry, more pressed for labor, and more likely to reforest some fields.

Plot forty-four was owned by a local man who had moved away. Since his death, his widow (who lives in Braga) has allowed the small dry plot to revert to native vegetation. Plot forty-five is a large plot with abundant stream water. It is now in wild grass, but the vines surrounding it are still tended. Because grapevines are a capital investment and will die within a few years without care, and are not much more difficult to tend in a small field than in a big one, vines may be tended even when other crops are not. Wine is also an emotionally important crop. In an early stage of reforestation, this plot is being gradually disintensified because it is distant and is made up of several narrow terraces.

Plot forty-six is an unusual case. The very large plot is divided into several terraces large enough to be mechanized, including several large fields. The land is at a high elevation, bordering the forest, with thin soil. Neighboring farmers report that the land is not quite as productive as fields at lower elevations in the center of the parish. The limitations of the thin soil of this farm are somewhat compensated by an abundant supply of water from the stream and from springs on the farm. The farm lies in the bottom of a deep stream valley. The neighboring farm across the stream—on very similar soil—has continued to be cultivated.

Soil and water conditions alone do not explain the decision to stop cultivating this farm. While about ninety-four hundred square meters are still planted in maize, the bulk of the farm is not even used for grazing. The farm has been surrendered to pioneer forest plants. Gorse and broom are growing in dense thickets in the fields. The grape arbors surrounding the fields have been abandoned, the vines are dying from lack of attention, and some of the supporting stone pillars have snapped under the weight of encroaching forest vegetation. The owning household consists of an elderly widow, her daughter, and the daughter's young children. The daughter's husband has worked in France since 1964, and she finds it difficult to keep up the farm and care for her children and mother.

This example is unlike previous cases, in which emigration was shown to play at best a supporting role in a few instances of land reforestation. This man has abandoned farming for his French construction job, even though the decision entails the loss of earlier, expensive capital investments such as grape arbors. This emigrant has earned the censure of the community. One parish resident spoke of the case for a long time, gradually growing angry as he discussed how gorse and blackberry bushes were taking over the farm that his uncle had once worked as a sharecropper. He concluded that "that animal has left a farm here to take a job in France, while some other man without a farm could have taken it!"

This case demonstrates problems in household economics related to macroeconomic changes. As this young man had an opportunity to work off-farm he declined to work land owned by his mother-in-law. The old woman cannot sell the land, because landownership is her status marker within the household, her source of security, and the legacy she will leave to her daughter in exchange for continued care in her old age.

Northwestern Portuguese of all socioeconomic levels and of all occupations place a high premium on keeping land under cultivation. Land that reverts to native vegetation or that is fallow is said to "grow old" (ficar a velha). To stop farming a large piece of good land is regarded as violating a public trust to use the land. Many people say that there should be a law to take land away from those who will not cultivate it and give it to someone who will.[2] Such sentiments apply much more to the case of this one large farm than to the other instances discussed, which are considered lamentable but largely excusable.

Until the midtwentieth century some marginal fields with little or no irrigation water were allowed to revert to native forest. This probably represented individual decisions at the household level to save labor. Reforestation may have reflected changes in household labor caused by death, emigration, and marriage. Plots with good supplies of water were never reforested. A good water supply ensured high yields and high returns to labor.

Very different kinds of fields were reforested after the midtwentieth century. When wages began to rise and agriculture became increasingly mechanized, small or terraced fields were reforested as well, even ones with plenty of water. Emigration was decisive in only one episode, which became a kind of local scandal.

All of the reforested land is marginal in one way or another: it either has thin soil, is distant, or is not mechanizable. The nearby fields in the center of Pedralva, with deep soils and abundant water, are never given over to native vegetation. There is nothing haphazard or sloppy about land reforestation: the decisions to reforest fields follow a clear, logical pattern. Those households facing greater labor constraints allow trees to grow on their plot with the lowest returns to labor, beginning with dry and distant fields. Narrow fields with gravity-fed irrigation water were among the most productive fields until tractors made them unworkable and water pumps made some early irrigation techniques obsolete—and the narrow, irrigated fields became the target of spontaneous reforestation late in the twentieth century.

NEW FIELDS

The reverse of land reforestation has also occurred as Pedralva's farmers have cleared forests for new fields. Table 5.4 lists the new fields in Pedralva and is organized like Tables 5.1–5.3, except that one variable (water) has been deleted—but is discussed in the text—while another (age) has been added.

The most noticeable aspect of Table 5.4 is the small number of fields, compared to the many reforested plots listed in Tables 5.1, 5.2, and 5.3. Reforested fields are easier to document than new fields. A field is like a building: it involves terracing, road building, and other land-altering work. Thus, a reforested field is like a ruin and can be observed on the ground, as an archaeological site can be. Reforested fields often bear agricultural features (wells, ponds, grape arbors, terraces, furrowed ground, large gates and large walls, and roads with turnouts for ox carts). Just walking in the woods, I found several reforested fields. Later I asked people about the history of the plots. Other people told me about reforested fields they remembered, as a

Table 5.4. New Fields

Plot	Owner	Amount of Cropland (in ha)	Age[a]	Years of Emigration
47	7	3.9	??	—
48	25	1.5	??	0
49	23	1.4	??	—
50	31	0.2	c 35	1960s
51	32	2.4	48	1957–1977
52	33	0.8	47	1959–1977
53	34	3.0	37	1970–1980
54	35	8.5	42	1962–1966
55	36	??	??	??
56	37	2.6	46	0
57	38	0	37	1969–
58	39	0?	??	now abroad
59	7	3.9	33	1970–1979
60	37	2.6	47	0
Mean				

[a]Age is of the current (male) farmer at the time of clearing.
[b]Added to 21,600 square meters of preexisting cropland to make a new field of 35,100 square meters.
[c]Added to plot forty-seven to make a new field of 17,740 square meters.

subtle way of criticizing their neighbors ("Just look at this land the farmers have allowed to grow old").

Which fields are new was more difficult to discover. Occasionally I found them by observing a field surrounded by immature grapevines (or no grapevines). In almost each case, the proud owner of the field confirmed to me that it was indeed newly cleared. Underreporting new fields may explain why there are fewer documented cases of new fields than of reforested land. Yet there is no doubt that since the 1980s the pace of land clearing has been brisk. Only eight cases of new field clearing are documented for before 1980. Of those, five plots (fifteen, sixteen, eighteen, twenty, and twenty-eight) were unsuccessful attempts to farm forestland (discussed in the last section) and have since been reforested. The three other cases are discussed below.

Plot forty-seven is a small field cleared by the great-grandfather of the current farmer. It was not irrigated but grew maize anyway, because some groundwater percolated out to the soil there. Land

Date of Reforestation	Size (in sq m)	Distance from Farmstead (in m)
early	3,410	600
1920s	2,000	1,600
1950s	few hundred	—
1980s	1,200	0
1981	5,800	450
1982	3,550	750
1982	6,800	650
1982	13,500[b]	550
1983	47,330	c 5,000
1983	6,150	0
1983	800	0
1984	3,760	0
1984	14,330[c]	600
1984	4,000	100
	8,090	860

cleared in the early twentieth century took a lot of work with hand tools and oxen to cut down the trees, remove the boulders and stumps, and level the earth. There has been more interest in land clearing since the 1980s when bulldozers became available.

Plot forty-eight is a series of small dry rye terraces at the edge of a larger maize field. It is owned by the same household as plot thirty-six, which was later reforested because it was small and distant, in spite of its good supply of water. Some households do both land re-foresting and land clearing, according to the microgeography of the plots concerned. Plot forty-eight was not reforested, because it is closer to home and is next to a large field, saving on travel time. Workers in the large field can tend the small terraces as well. Plot forty-nine is also a small terrace, next to a preexisting field, originally cleared by a household in a nearby parish.

The remaining plots, cleared between 1980 and 1984, all belong either to emigrants or to large farmers (or to large farmers who have been emigrants). Their ages range from thirty-three to forty-eight, suggesting that it is the ambitious young farmers who are interested in expanded their farm size through forest clearing.

Land clearing is a type of technical change that involves relatively little capital investment (Bentley 1987b). Farmers claim that selling off timber nets enough cash to pay a contractor to clear and level the tract, and to have a well dug and an irrigation tank built. The farm family spends slack days plowing the newly cleared land one row at a time, unearthing large rocks that they carry to the side of the field and use to build a wall around it.

Plots fifty-one to fifty-three, fifty-six, and sixty represent significant additions to middle-range farms. Most of these farmers have been emigrants. These people tend to be ambitious farmers, with money to invest in machinery and a strong desire to expand farm size. Plots fifty-one, fifty-six, and sixty have water from a newly dug well on nearby forestland.

Plots fifty-four and fifty-nine are expansions of existing fields by larger farmers. Plot fifty-four was added on to four preexisting smaller fields to make one new, large field. It is watered from an existing well. Plot fifty-nine was added on to a preexisting smaller field. The owners are building a modern cow barn on their farmstead and plan to expand herd size with maize silage from the new field. They will irrigate the new field from a well and tank at the farmstead, moving the water by gravity flow to the plot. The liquid manure from the barn's septic tank will flow down a pipe directly to the field, watering and fertilizing it while saving transport costs for manure, the heaviest agricultural input, or supply.

Plot fifty-five is an anomaly; the owner is a young farmer in a nearby parish, whose wife is from Pedralva. The plot was originally moorland, not forest, so there were no trees to sell to defray land-clearing costs. The farmer borrowed money from his father-in-law to have the large area cleared. He did not have the capital to farm more than a little of the area. The soil is very thin and stony, although there is water from a shallow well. In 1984 the owner allowed poor people from Pedralva to plant 7,520 square meters (16%) in potatoes, just to keep the land from reverting to scrub and to fertilize the soil. He also planted maize on 7,140 square meters (15%). He cut the maize for silage and hauled it about five kilometers to his home, where he kept his cows at his father's farm.

The new fields are all fairly large, especially the ones created by farmers (as opposed to fields created by emigrants). Most have irrigation water, often from a newly dug well. Water is frequently moved to the new fields by electric or gasoline-powered pumps. The adoption of water pumps in the 1940s allowed farmers to clear forestland that they could not have irrigated before. New fields tend to be fairly large and close to the farmstead because farmers consider size and distance while deciding which forests to fashion into fields. The increased supply of manure from dairy production (after 1978), mechanized transport for hauling manure, and chemical fertilizer allow farmers to convert even poor soil to fertile farmland.

Forest management—specifically reforestation and clearing for fields—may appear idiosyncratic and random unless seen in the context of the total farm economy. Decisions to change a site's use are influenced by the land supply and by technical changes in agriculture. These decisions are often the result of household decisions to adjust land to labor and capital supplies.

The patterns of reforestation and forest clearing are geographical (certains kinds of land are reforested or cleared) and historical (technical change through time affects which types of land are chosen for change). Because large and small owners live in different economic environments they are not likely to make the same kinds of changes. Land use changes occurring over different periods of time, and brought about by landowners of different farm sizes, are patterned adaptations to the natural and economic environment.

6. The Earth Has Bones

On a fall day in 1984 I sat with an elderly farmer, watching a bulldozer rip boulders from what was left of a forest plot, converting it to a field. As the machine labored with a particularly large stone, the old man turned and said, "a terra tem ossos" (the earth has bones). The analogy of the earth as a living being is reflected in other things people said, such as "water is the blood of the earth." The Portuguese farmers have a profound respect for, and understanding of, their land. It is a tribute to their thoughtfulness and farming skills that they have changed their style of farming so much in the past two decades while preserving the beauty and value of their land. But while the pre-1964 style of farming was sustainable and ecologically sound, people went hungry. The modern technology uses some agrochemicals and has more of a market orientation, and is probably not as acceptable to agroecologists, but the contemporary style of farming is more satisfying economically to those who do it (see Richards 1989a on agriculture as performance).

Farming behavior is formed by the interplay of culture and individual invention with the economic and natural environments. The people of Pedralva share a common culture but adapt to their environment through different behaviors. Each household lives in a unique natural and economic environment, and the agronomic behavior of each household is correlated with farm size. Small farmers have scarce land and capital but abundant labor. They are reluctant to allow even marginal fields to revert to forest and do not mind long walks to far-flung parcels. They spend as little money as possible on seeds, fertilizer, and other purchased inputs. They tend to use hand tools in place of machinery, and they use hoes to work the field corners and small fields that tractors cannot reach.

In Chapter 2 the smaller farms were shown to use more labor-intensive techniques and less machinery than larger farms. Knowledge of how to cultivate plants and tend livestock is cultural lore that is shared by most community members, even nonfarmers. How people behave in their gardens and fields depends not only on the common culture they all share, but on their different supplies of land, labor, and capital. Machinery is a capital-intensive solution to a labor shortage, while hand tools are a labor-intensive solution to land and capital constraints. Smaller farmers do not avoid using machinery because they are closer to the earth, because they are inherently traditional, or because they are women who do not often visit extension agents. They substitute labor for machinery (capital) because they have more labor than money.

Everyone in the community knows many agricultural techniques and there are several ways to do any given task. For example, maize can be planted by hand or in a seeder pulled by a cow, a horse, or a tractor. Mere knowledge is not the key to determining which technique a household uses. People try to maximize the efficiency of (returns to) the scarcest resource. A person with two hectares of land and little outside labor has half the labor per unit of land as the person with one hectare. People with more land have more cows and higher farm profits. So, as a general rule, the more land a household has the more capital it has. The larger a farm is the more the farm family attempts to save labor by substituting capital equipment for labor. The smaller farms must save capital, so they use more hand labor and less machinery and buy fewer seeds and chemicals.

There is no single technology or behavior for a community. There is a complex set of technologies that people choose from to make the best use of their resources, that is, the closest fit to their personal economic and natural environment. Larger farmers tend to form work parties for some tasks, such as potato harvesting and rye threshing, that require many hands. The workers are often not paid but are treated to festive meals and snacks. Pedralvans say rhetorically that they "help their friends, as a favor." Helping someone by giving labor implies that the favor will be repaid. Farms of about the same size tend to repay labor with labor, and machinery use with machinery use. Smaller farmers may give labor to larger producers in exchange for machinery use (for example, the larger farmer may come plow a field for the smaller one). Often no exact exchange is specified, so smaller farmers and nonfarmers help larger farmers but receive nothing in return at that time, although the larger farmer owes them a favor that is generally repaid at some later time. This is a form of generalized reciprocity, and people who exchange favors are involved in long-term exchange relationships, in which the exchange is not supposed to be too closely calculated. They say, "não fazemos contas" (we do not keep track), meaning that they do not keep a tally of the exchange of goods and services, only a general reckoning as to whether the exchange is more or less equal, and in good faith.[1] Some larger farmers give away something they do not need and cannot use, such as surplus irrigation water, that a neighbor needs desperately. The smaller farmer who receives an irrigation turn "as a favor" may be obliged to repay the favor in labor, by working for the larger farmer. Since the deal is never formalized as a trade, the smaller farmer accepts the water turn as a favor but then works for the larger farmer whenever the larger farmer asks for help.

The tendency for larger farmers to use machinery and other innovations was given a historical perspective in Chapter 3. Larger farmers are consistently the early innovators, often substituting capital for labor by buying (or renting in) machinery before the smaller farmers make the same substitution. Before the period of heavy emigration (beginning in 1964) the larger farmers used more nonhousehold labor than smaller farmers. After emigration larger farmers substituted capital equipment for much of the labor of the landless who left Portugal. However, although the smallest farmers face the greatest land constraint, they are not the first to experiment with such land-saving inputs as chemical fertilizer. New breeds and chemicals invariably cost money, which is usually in short supply on small farms. Although fertilizer is the classic land-saving input, I also consider it to be a labor-saving device in the Minho, where it replaces the labor of brush cutting (Silva 1983; Caldas 1981). Because of the "facilitating effect of wealth" (Cancian 1979) larger farmers are the first to try land-saving innovations, but smaller farmers generally follow suit if they have money on hand to buy the new inputs. The most important agronomic innovation is the recent introduction of dairy cattle, which spread to all but a very few farms between 1978 and 1984.

Information, or knowledge, is not the most important variable in the choice of farm technology. While large farmers occasionally go to town and visit extension agents, this knowledge is rapidly diffused through the community. The first farm to instal a milking machine learned about it from extension agents, but other villagers learned about the machine from the local innovators. When the machine was first used the neighbors gathered around to see it work, and when they learned how profitable milk was they quickly developed an interest in commercial dairy production. The high profits were founded largely on state subsidies to raw milk prices.

The path of technical change is from large farmers to small farmers, who may lag behind large farmers for several years before eventually changing. Large farmers are allowed to assume the initial risks and to experiment with the technology, adapting it to local conditions. By the time small farmers adopt an innovation it does not represent a risk to them; they have seen it work for several years on the larger farms. For example, when silo technology was first acquired from extension agents in the early 1980s farmers built silos just as they were instructed by extension agents—deep concrete-lined trenches with a high gabled roof. Farmers soon learned that silos could be built with less-expensive material. Farmers who built silos in the mid-1980s saved a lot of money by building them according to a simpler design that was the result of local experience.

There are several sources for innovations. Some innovations come from extension agents, as commercial dairy production did. Others come from neighboring parishes and from agricultural input merchants (seed salesmen). One large farm has been the first to try most innovations. Other innovations have been tried first by a handful of the largest farms. Small farmers, including women, are not denied access to new knowledge just because they are less likely to visit extension agents. Knowledge learned from extension agents or any other outside source spreads rapidly through the community. Smaller farmers who exchange labor with larger farmers learn new techniques from them. Only a handful of large farmers visit the extension agents, but then relatively little new information is learned from extension agents, and any information that is so acquired is accepted cautiously. As one man said, "Why should we listen to them when they don't know the difference between a cow and a calf?" He was referring to the fact that few extension agents have a farm background, and their education is largely in the classroom.

Larger farms have larger fields, but the other indices of land fragmentation are more pronounced on larger farms than on smaller farms. The insignificant correlation of farm output and fragmentation suggests that land fragmentation is no barrier to agricultural production. Conventional wisdom has it that land fragmentation leads to inefficient use of land and labor. Portuguese agronomists and policymakers in particular blame land fragmentation for a host of the country's agricultural ills. Economists and geographers point out that fragmented farms increase travel costs to farmers and do not allow efficient machinery use. Anthropologists have discussed the ecological and economic benefits of land fragmentation; having multiple plots of land on each farm helps to minimize risk and take advantage of different natural environmental zones, and allows for better crop scheduling.

Unlike other communities where anthropologists have studied land fragmentation, there is little environmental variation in Pedralva, so land fragmentation does not offer a strong ecological advantage. Land fragmentation in the Minho is ecologically neutral, neither increasing nor decreasing milk production. Land fragmentation is socially adaptive, allowing farmers to fine-tune their land-to-labor ratio, so that total usable farm size can change gradually with the household's developmental cycle.

Land is fragmented in all areas of the world that practice intensive agriculture. It is caused by land transfers (sales, rent, inheritance) and by clearing fields from the forest in small parcels. Land fragmentation is an adaptive response of intensive cultivators to conditions of scarce land, and often to unique, local natural environmental conditions. Fragmented land does not represent a backward or irrational agricultural structure. Land consolidation is beneficial only in countries where the rural population is rapidly leaving agriculture, so the land structure must be quickly adapted to the rapidly diminishing labor supply and, generally, an increased supply of capital. Because land consolidation is expensive it is an option only for wealthy countries with public funds to buy land and administer land exchanges (Bentley 1987a).

Large farmers in Pedralva clear some parcels and reforest others to adjust their land and labor supplies under conditions of changing farm technology and changes in the relative supplies of land, labor, and capital. Large farmers and small farmers have different attitudes about land use, stemming from their unequal access to land. Larger farmers are more likely to complain about land fragmentation because they have more fields to visit. Smaller farmers and nonfarmers point to reforestation as an example of larger farmers' antisocial behavior. Land is a public good in a sense, and farmers are expected to be stewards of it. Even though land is owned privately, there is a notion that people who own it should be obliged to work it to produce food for the people. Farmers who allow fields to go out of food production risk being criticized behind their backs by others, even though they regard the decision as rational.

The household supplies of land, labor, and capital determine the decisions people make about what technology to use, and whether to use a given piece of land as forest or field, in spite of the fact that all community members have a common stock of knowledge about land and technology. Political power is not entirely in the hands of the wealthy. The current president of the Parish Council (Junta da Freguesia) is a retired laborer, while the previous one was a wealthy farmer. The Parish Council built a large town hall (sede da freguesia) in 1984, in spite of the farmers' complaints that the highest municipal priority should be cobbling the many village roads (for improved access to their fields). In 1986 the Junta cooperated with the Câmara (chamber, specifically the local government building, generally understood as the borough government itself) to have the dirt roads cobbled.

Farming behavior is not the only aspect of life that is affected by socioeconomic stratification. Nonfarmers complain behind the farmers' backs about how the farmers are stingy and greedy. People say that farmers want them to work for nothing. The farmers complain, in their turn, that nonfarmers are lazy and that no one wants to work anymore. Large farmers say that small farmers are not technologically progressive because they are uneducated. Wealthier farmers complain that it is unfair for returned emigrants (generally former pobres) to have foreign currency bank accounts: they are "full of money"; their large tile-covered houses are ugly and pretentious; they no longer have respect for others because they think that their money has made them lords (fidalgos).

Although the community is divided along socioeconomic lines, there are many links between people of different economic positions. Even pobres, who criticize farmers in general, have ties with some individual farmers and work for them as part of an institutionalized friendship. Farmers who criticize emigrants have friends and kin who are emigrants, and may have once been emigrants themselves. Farmers sometimes complain that "no one wants to work anymore" even when surrounded by a party of nonfarmers working for them, without salary.

I have relied heavily on the contrast of farmers and nonfarmers, and of large and small farmers. Large and small farmers are different in several ways. The large farmers tend to be men and women who invest more money in their farms, because they have more capital and land and less labor than smaller farmers. Smaller farmers tend to be women whose male kinsmen work mainly off-farm. They save their scarce capital by lavishing labor on their small fields. Larger farmers have more fragmented farms than smaller farmers and tend to obtain their farms through inheritance, while smaller farmers get their land through a mixture of inheritance, rental, and purchase.

In personal terms, the darker side of socioeconomic stratification in Pedralva was not that the poor were held to be socially inferior to their neighbors, but that they did not have enough resources to live the life that they felt was minimally acceptable. Local people claim that the only people who really live well are the larger farmers. Before emigration began in the mid-1960s, the nonfarmers did not always earn enough to feed their children, who were sometimes given to farmers as field servants, to work for their meals and a homespun linen shirt every year or two, often sleeping in the hayloft. They would be sent on such lonely tasks as goatherding, so they often missed school. Some people say that their parents nearly starved to death trying to keep the children at home. After emigration began, near-starvation poverty was traded for the pain of family separation, as men left their families to work in France. Families that have sent members abroad never quite recover from the emotional shock of the long years of separation. Husbands and wives miss each other, children grow up without their fathers. Women must assume heavier farm tasks. The men in France often live in a barracks or crowded apartment, working and keeping house for themselves for eleven months of the year in a hostile social environment, merely to support the families they seldom see. Of those who stay in Portugal, life is a lot of hard work for everyone, and few luxuries for all but the largest farmers. People complain constantly about their lives: the work is too hard, too dirty, and they do not earn enough money.

Pedralva's social structure incorporates some aspects of both gemeinschaft and gesellschaft. The wealthy peasant is distinguished economically from the poor peasant, but they are both country people and therefore considered inferior by the elite and middle class of their own nation's cities. The social distance between wealthy peasant and rural poor is not as great as the social distance between rural people in general and the urban middle class, but it seems greater than Pina-Cabral (1986) suggests.

Only a few of the large farmers are still technically peasants, living off the land and selling only a little farm produce. The small farmers and nonfarmers with large gardens live by a mixed strategy of subsistence farming and working for a wage in town or abroad, while most of the large farmers are becoming commercial growers. Since the mid-1960s Pedralva has become a suburbanized community of worker-peasants on the periphery of Europe. The rural population is no longer predominantly agricultural, as many people, especially men, now work in blue-collar trades, locally and in France. By diversifying their economic base, worker-peasants are able to enjoy a higher standard of living than their parents did, and a higher standard of living than contemporary urban workers. The cash income of the typical construction worker is only about Esc 20,000 (1984) a month, about U.S. $125, a difficult wage to live on without the women's contribution of garden produce and rabbits and chickens. Women farmers buy copper sulfate fungicide for grapevines, and other farm supplies, with money that their blue-collar husbands and sons bring into the household.

Other benefits of the suburbanized countryside include the possibility of inheriting a house, cottage, or lot. The country dweller can build a house over several years by working on days off with the help of friends. By not moving to the city, villagers easily maintain kin ties. It means a great deal to men to be able to play cards with their boyhood friends in the tavern on Sunday, just as it is important to women to visit their mothers and sisters as they do the wash or work in their gardens.

Most of the men of Pedralva now work in town. By keeping their rural homes, elderly household members and women with dependents have productive, convenient agricultural work. They have less-expensive housing and keep kin and friendship ties. The modest standard of living they enjoy is due especially to their inventiveness, frugality, and hard work, as well as to political and economic conditions in the world outside—over which they have little control—and to the economic rationality that they bring to lifelong decision making. Adopting chemical fertilizers, dairy production, tractors, and other innovations has helped make agriculture more profitable, if not more ecologically sound, increasing farmers' incomes. With certain notable exceptions, such as farmers' complaining about increased wages, there is a general agreement among Pedralva's people that though life is hard now, it used to be much worse, so much so that today there is no misery.

7. The Future of Small-Scale Farming in Northwest Portugal

I have described Pedralva's agriculture as eminently logical, with rational economizing embodied in local practice and decision making. The agricultural system has great internal variation, is highly evolved, and well adapted to the local natural and economic environment. This view is at odds with current Portuguese government and foreign economists' views on the subject. According to received wisdom, northwest Portuguese agriculture is stagnant, backward, and inefficient (Pearson and Monke 1987:17). At least one anthropologist, in an otherwise excellent ethnography, used the phrase "poor technology" to characterize the agriculture of nearby Trás-os-Montes (O'Neill 1987a:125, 297).[1] What does the future hold? Indeed, will there be a future, or will small farmers simply abandon farming as some experts expect them to do?

During my field work I was associated with a team of U.S. and European economists. Since the team's conclusions have recently been published (Pearson et al. 1987) it is appropriate to address some of the questions they sought to answer. Anthropologists use a more fine-grained, local approach than economists, often living in and observing one community. But anthropologists also try to identify with the people they study, and try to explain the people's perspective rather than assuming that one set of economic rules can account for their behavior. The ways in which my findings support or contradict the conclusions and policy recommendations of the economists stem largely from the difference in our perspectives.

The approach taken by Pearson and other members of the team is innovative. They introduce the notion of private and social (public) profitability, which allows them to discuss how a farmer could be making a profit (private profitability), but at a high cost of government subsidies (rendering it not socially profitable). They have also gone to a great deal of painstaking work to show how the private and social profitability of important commodities will change as Portugal adapts to European Community (EC) conditions. Their work is explicitly predictive, a tack that anthropologists do not often take. That I disagree with some of their conclusions is in no way a negative reflection on the caliber of their work. Anthropologists and economists have trained themselves to see the world in remarkably different ways.

LAND FRAGMENTATION

Economists generally see land fragmentation as an agricultural problem (Monke 1987a). Both in this book and elsewhere (Bentley 1987a) I have argued that land consolidation schemes are generally not worth the expense. A fragmented land structure is the end result of a long period of evolution and has certain economic advantages over farms concentrated in single plots. Policymakers consistently express strong opinions in favor of land consolidation, even though there is no evidence to support the cost-effectiveness of such schemes. Perhaps the irrational drive to consolidate land should be seen as an expression of a cultural value of modernism, a desire to attain the symbols of a "modern, progressive" agriculture, which is believed to be so inherently superior to traditional agriculture that little analysis is needed to justify sweeping agrarian changes. Whether the changes are productive or not is less important to policymakers than becoming "up to date." Farmers' objections are simply dismissed as irrational, conservative, and traditional.

Land consolidation processes begun in 1980 focused on three concelhos in Portugal, including Braga. Only 450 hectares were consolidated—enough to make up one good-sized family farm in the United States. Monke (1987a:75–76) concludes that efforts were hampered by the lack of cadastral surveys and by limited funds. It is no wonder that funds were limited, since a scheme in Beira and the Algarve between 1978 and 1983 consolidated a scant 750 hectares, but at a cost of Esc 1 million (U.S. $11,100 in 1983) per hectare. As always, most of this expense was in public sector salaries. Since thirty technicians labored for five years to consolidate 750 hectares, the net result is a mere five hectares consolidated per technician per year.

Monke's (1987b:254) data show that parcel size in northwest Portugal is increasing through spontaneous consolidation. He also shows an increase in the number of very small farms (under four hectares) and very large farms (over a hundred hectares). Monke writes that government policy should simply remove restrictions on the land market to allow this natural process to continue. This is a sensible approach, if not for its belief in the magic market then at least for the notion of not meddling in the existing land structure; I hope policymakers will heed it.

THE GROWING IMPORTANCE OF WORKER-PEASANTS

Economists and policymakers have a bias in favor of big farmers, or a prejudice against small farmers. Official policy discriminates against the rural poor. As Fox and Finan (1987b:219) comment, the task of extending new technology to small farmers "is made even more difficult by the limitations of certain EC structural aids that exclude small farms."

Economists and Portuguese policymakers do not envision increasing numbers of small farms, part-time farmers, semiretired farmers, and returned emigrants turned small-scale farmers in the agriculture of the future. Pearson and Monke (1987:21) lightly dismiss more than half of Portugal's 900,000 farms because they are less than one hectare in size, and because "few of these farms depend on agriculture for a significant share of family income."

The negative view of small farming ignores economic as well as cultural issues. A household with half a hectare of land regards that plot as an extremely important part of its portfolio. People with dependency burdens and other chores to do (especially women and the elderly) cannot easily take off-farm employment, but they can spend a few hours a day working on the land to supplement the household food supply, especially by growing good, fresh produce. A small plot of land is the safest form of savings, especially since real rates of interest are nearly nothing. The lot also is a potential site for grown children to build a house of their own. Because small farmers can apply a lot of labor to their tiny farms, their land tends to be as productive as any.

Economic benefits notwithstanding, important cultural values and motivations lead villagers of northwest Portugal to emigrate, endure years of sacrifice, and return home to buy a little patch of land. Having land is a symbol of wealth and status. Since many, if not most, people in the region are born without land, being able to buy a nice parcel and build a house on it are real accomplishments. Men who migrated to France without their families in the 1960s generally worked in construction. After twenty years abroad they often had high-paying jobs, especially those who had learned to speak French. For these people, coming back to Portugal was a decision to sacrifice economic motivations for family and home. The land they bought with money saved in France is more than security and a status symbol. One of the men I knew in Pedralva was just retiring from his job in France. He was spending his first spring home in twenty years planting potatoes and vegetables around his house. The little garden had rich black earth, mature grapevines all around it, and a stone wall to keep out neighbors' animals. The family home was comfortable, old but remodeled. With obvious pride the man explained, "I was born with only two things, the sun and the moon." Little can match the sense of accomplishment these people feel when they have a nice house and garden. They started life with no land, no money, and no chance for an education, and everything they own is the fruit of their hard work. Having land allows one to be one's own boss, to work in the out of doors, to follow an occupation they know well. It also allows people to spend time with their children, working in the garden and teaching them about crops and animals .

Part of the problem economists have in accepting very small-scale farming as a legitimate activity is that the smallest of the small farmers are generally worker-peasants, a group that does not dissolve easily into the categories of formal economics. The worker-peasant phenomenon is not new in Europe (Holmes 1983), and it is not going away. It is increasing both in terms of number of people and importance. The woman and children left behind by a husband in France, and the retired emigrant family with its one-cow farm are worker-peasants. So is the Portuguese construction worker whose job is in the city but who takes the bus home every night to the village of his birth. The lunch his wife packs him may include potatoes, eggs, kale, and wine produced on their own land. The women in such households work on the land daily, while the men provide extra hands on Saturdays and holidays for harvesting, planting, and other chores that take a lot of labor. Since worker-peasant households sell little, if any, of their crop, the man's salary can be used to buy the fertilizer and other supplies for the garden. Certainly in Pedralva worker-peasants

are the largest socioeconomic group. Extension agents familiar with
the whole Minho told me that in some other parishes there are no
large farms and almost everyone is a worker-peasant.

THE ROLE OF SMALL FARM

While it is no longer fashionable among economists to characterize
peasant farmers as irrational, it is still possible to describe their farming as unprofitable (and so by implication irrational). Economists
complain that worker-peasant farming in the North, on one's own
small farm, does "not offer returns sufficient to cover the market
value of labor" (Pearson and Monke 1987:23). This reductionist perspective assumes that people are always able to find employment,
which they are not. It also assumes that salary is the only considera-
tion in deciding to stay home and work on one's own small farm or
take the bus to Braga every day to carry hod or clean toilets. Many
people are locked out of the off-farm job market, especially women
and the elderly. Portuguese villagers remain economically active
until they are too feeble to work. Eighty-year-old people are often
seen herding cattle, harvesting potatoes, husking maize, and doing
other useful tasks. Women with small children or elderly parents to
care for are not able to work off-farm if there is no one to care for
their family. Agricultural chores can be carried out in between household tasks. Women without dependency burdens may work off-farm
if they are willing to take the low-paying jobs open to them in the
service sector.

Portuguese villagers will work for others, but all other things
being equal, they prefer to be their own bosses. Working at home on
family land allows people to be productive but to work their own
hours, at their own pace, and without being told what to do. The
work is secure and may be more productive than working off-farm.

In some ways economists and policymakers are like the hedgehog
who knows one big idea, but knows it well. In Portugal they place
almost all the emphasis on increasing farmers' incomes. Since they
see no way to raise returns to labor, and since existing farm subsidies
will be phased out by 1996 to match those of the EC, they foresee a
drastic decline in the profitability of certain crops, especially milk.
Some economists and policymakers conclude that the only way to
raise farm incomes is to dispossess most of the farmers! One plan by
Fox and Finan (1987b:211) suggests "displacing" 300,000 people

from small farms in northwest Portugal, to make room for big dairy farms: "if the approximately fifty thousand milk producing families who would be displaced by the medium milk structure had six members each, 300,000 people would be affected. Assuming that one-half of these people stayed in the rural areas to provide labor on the larger milk farms, to continue farming, or to retire, migration or off-farm employment would have to support 150,000 people." In other words, half of these 300,000 small farmers should retire or become peons on the big farms, while the other half should take (as of yet nonexistent) factory jobs or leave their country. Fox and Finan then go on to suggest that up to half a million people could be disenfranchised in this way from wine-producing farms.

These notions are consistent with government schemes. The Beira Litoral program seeks to reduce the number of farms from 30,000 to 5,625 in fifteen years. One project for modernization of vinho verde production in the Vale do Lima calls for reducing wine producers from 37,443 to 12,355, of which only 1,125 producers would benefit, since they would have three thousand of the forty-seven hundred hectares. The rest (11,230) would use seventeen hundred hectares for home wine production (Fox and Finan 1987b). Even Fox and Finan realize the nearsightedness of their own ideas: "although these programs appear to be technologically sound for the project beneficiaries, they ignore the farm families that would be displaced by the creation of larger farms. The displaced families are relegated implicitly to the category of social problems that are not considered to be the responsibility of the Ministry of Agriculture and the Regional Agricultural Services. Their problems are to be solved by other ministries and agencies" (p. 218).

This is like suggesting that small farmers should be thrown off their land and given food stamps so that their wealthier neighbors can afford to take their vacations in the Algarve. The birthright, livelihood, and way of life of the people of an entire region is considered unimportant compared to increasing the income of the largest farmers in that region. The worker-peasants and small farmers who would be displaced by such a scheme have no real alternatives. If there were high-paying industrial jobs in the region more people would take them. Since there are not enough local job opportunities now for the people who would leave the agricultural sector if they had a chance, displacing villagers from farming by fiat would only create a large impoverished, dispossessed group of miserable people. They could take up some of the marginal, underemployment kinds of activities already seen in Portuguese cities—begging, selling plastic toys, shining shoes, and similar tasks. Another option would be to create a peasant underclass, dependent on welfare, transforming villages into rural slums.

Yet another option would be to increase emigration. Because northern smallholders who work abroad often save their money in Portuguese banks, emigration would increase the supply of capital for southern industrialists and would relieve Portugal of the task of finding employment for its own people. This "solution" ignores the reality of contemporary migration on at least two counts. Portuguese who have been out of their country (especially in France) for years are returning home, citing unemployment, bigotry directed at foreigners, and the lack of employment opportunities for their children as the reasons for returning. Planners who glibly suggest emigration as the panacea to Portugal's employment problems ignore the human cost. They never mention the pain and confusion that Portuguese villagers experience while working abroad. The experience is hardest on men who leave their families in Portugal. Issues such as a fulfilling family and community life and job satisfaction cannot be discussed entirely within an economic paradigm, with its emphasis on numbers and money. One man I knew in Pedralva had neither house nor land, but had a growing family. In 1970 he emigrated and was gone for fourteen years, coming home only for his August vacations. A few months after he returned for good in 1984 the subject of his French pension came up in casual conversation. Tears welled up in his eyes and his voice cracked as he said, "Don't you think I deserve it, after all those years I spent in France?" Like people anywhere, the Portuguese country folk love their families, and the many years they spend apart are quite painful for the men alone, the women left with the land, children, and no husband, and the youngsters whose father is always away. It is absurd to think of driving more people into foreign exile as creating opportunity for them. It is certainly not the most humane way to free up more land to increase farm income for what is already the wealthiest sector of the region: the large farmers.

DARK DAYS AHEAD?

The future of northwest Portuguese agriculture is probably not as bleak as the economists' predictions, based on the assumption that prices can be predicted into the future and that there will be no significant change in technology. We have seen in Chapter 3 that technology has been changing rapidly for forty years, and very rapidly for twenty. Pearson and his colleagues predict that because milk subsidies will be phased out, dairy production will change from a highly profitable activity in 1984 to a highly unprofitable activity by 1996. An alternative would be to raise more potatoes, but that is assumed to be impossible because of "disease constraints." Economists assume that farming will become unprofitable in the Northwest and that farmers will leave farming because nothing can replace the profitable dairy sector, soon to become unprofitable. Neither a new technology nor expanding the existing potato production is seen to offer any hope. The economists' dire prediction of total collapse of the small farm can be turned on its head simply by demonstrating that Minhotos could grow potatoes.

The assumption that potato production cannot expand (Monke 1987b:248; Pearson 1987:265) is based on Finan's (1987:151–52, 163) simple assertion that "potatoes are very susceptible to a wide array of soil-borne diseases and cannot be produced safely on the same area over consecutive years. Farmers acknowledge the problem of potential falling yields and move their potato plantings from field to field" (pp. 151–52). This observation is based on neither long-term field work with farmers nor on plant pathology research. Because farmers told him that potatoes cannot be produced in the same field for many years in a row, Finan assumed that farmers are already growing as many potatoes as they can. Finan presumes that one-tenth of a one-hectare farm is devoted to growing potatoes (Fox and Finan 1987a:193). Thus, according to Finan's logic, although a farmer could move the potato patch every year for ten years without putting it in the same place, disease risk keeps farmers from planting more potatoes.

I observed farmers in Pedralva plant large fields of potatoes for two years in a row before moving them. Cottagers with small gardens (less than a thousand sq m) plant one-half of the garden in potatoes and the other half in vegetables. The following year they move the potatoes to where the vegetables were. They are able to keep about half of their land in potatoes every summer. Clearly there is need for some basic agronomic research in this topic, but, assuming that farmers can grow a crop of potatoes for two years and must then grow maize or another crop on that land for the following four years, a third of the farmland of the Minho could produce potatoes every year. Even accepting a conservative estimate of 20 percent allows for a doubling of potato production (accepting Finan's estimate that 10% of the area of a "typical" small farm is planted in potatoes).

Finan (1987:152) confusedly claims that because potatoes have to be moved from field to field that "the small size of northwestern farms . . . poses a constraint to significant expansion of potato production." Having a big farm allows farmers to grow more potatoes, but it does not help them devote a higher proportion of their land area to potatoes. My impression is that the small, fragmented Minhoto farms are ideal for avoiding the problems of soil-borne potato diseases. Potatoes were rarely planted in patches larger than two thousand square meters (the largest ever in Pedralva was only five thousand sq m). The patch was generally placed in the corner of a field, further fragmenting the fields through multiple crop regimes in a single parcel—to no apparent detriment to crop or farmer. Sometimes a very tiny field would be planted entirely in potatoes (intercropped with kale and surrounded by grapevines). Having several fields allows farmers to plant more patches of potatoes, isolating each patch and reducing the risk of total disease loss. The existing farm structure in the Minho eases the future expansion of potatoes rather than restricting it.

Economists also forecast the demise of northwestern agriculture on the assumption that technology will not change. In part this is due to a lack of confidence in the Ministry of Agriculture to develop technical change through strategic research. It is also based on an unjustified assumption that technology does not change through local adaptation and invention. In this book I have mentioned a new planting style for maize and beans invented by a local farmer. The new potato technology used in Pedralva was picked up from farmers in Trás-os-Montes by a local farmer who went there to learn it. The production of broadcast forage maize was also presented to me as a local invention.

While I was in the field I visited several other parishes near Barcelos where an extension agent was successfully introducing greenhouses to rural women with very little land. Simple structures of plastic sheeting and some wood, the greenhouses produced tremendous quantities of vegetables and fresh flowers for market. The women who operated them told me they were completely satisfied with the greenhouses.

The doctor from Pedralva, who manages his relatively large farm on a part-time basis, suggested half-jokingly that he owned so many ruined stone houses (formerly for sharecroppers) that with such low labor costs he could easily grow mushrooms in his old houses. I am not suggesting that greenhouses and mushrooms will save the agrarian economy of the Minho, but neither am I ruling out the possibility of creative, important change. The future cannot be predicted with the precision that Pearson and his fellows think is possible. Their unrealistic assumption that technology is static is necessary for projecting prices into the future. The adoption of some new crops or crop mixes or new technology could greatly increase returns to land and absorb more labor.

The villagers of the Minho are not passive victims of the world system, about to be crushed by Portuguese accession to the EC. We should listen to what they want: an increase in income, while remaining in their native villages, with no objection to taking industrial employment. Small factories opening shop in Minhoto towns would find a disciplined, hardworking labor force, willing to work in industry while continuing to tend their small plots.

We should also pay attention to what the people's behavior consistently demonstrates. They are hard workers unafraid of change, emigrating and taking up new work patterns, adopting new crops, animals, machines, and chemicals. Because they are a highly successful, adaptive people, the rural northwestern villagers should not be seen as anachronistic European hillbillies living in the twilight of a fading era. They will adapt to decreases in milk prices by changing their technology and their crop mix, although the details of that change are not predictable. If hard times come for the Minho it will not be the first time that the Minhotos have experienced the oppression of poverty. It is not realistic, based on either past experience or present behavior and attitudes, to expect smallholders to adapt to a farm depression by packing up and leaving their land for the wealthy farmers. Land, even a little, is still the most valuable asset in the Minho, and poverty induces people to make more-intensive use of it, not less.

Pearson (1987) suggests that the predicted farm crisis in the Northwest needs to be anticipated by government programs aimed at improving and extending agricultural technology, and by programs to provide job information, retraining, extended unemployment insurance, and long-term education for rural families. While this is a humane, thoughtful approach, it ignores the fact that no amount of job training can help underemployed villagers find jobs unless new jobs are created. Government policy should work to encourage a dispersed, rural industrial base. Small labor-intensive factories in the countryside would provide the kinds of jobs people want, in the communities where they want to live, and take the sting out of falling farm prices.

Agricultural planners should expect that rural workers will hold on to their large gardens and small farms. Such a development should be greeted as a successful adaptation and not as a counterproductive medieval survival. Like a fragmented farm with several parcels of land, a household that puts some of its labor into salaried jobs and some of it into farming successfully manages risk. The industrial job provides money and protects the family from total crop failure. The land guards against unemployment, supplements the diet, and absorbs labor that cannot be sold on the market.

Government policy and economic theory should recognize the social benefit of a worker-peasant adaptation. In fact, worker-peasants are conserving both the natural and the human resources of a countryside, keeping it productive and satisfying what are apparently deeply felt social and psychological needs.

Notes

CHAPTER 1. A PARISH IN THE MINHO

1. By suburbanized I mean that many of Pedralva's residents now work outside of the parish but commute home every evening, or in the case of foreign migrants return once a year for a summer visit; no urban people have yet moved to Pedralva to build country homes and commute to work in the true suburban sense.

2. For the Portuguese version of this article see Iturra 1983.

3. Following Pina-Cabral's (1986:1) translation of the term.

4. This estimate of ten square kilometers was taken from maps. The land surveyors' total of 7.27 square kilometers includes only fields and forests. The surveyors' data were measured on the ground, and the cumulative error of taking many quick measurements from over sixteen hundred irregularly shaped plots is difficult to judge. Much of the remaining 2.73 square kilometer discrepancy is no doubt taken up by a highway, many roads and paths, over three hundred houses, several ruined houses, two chapels, a cemetery, a town hall, school, and about eighty barns and threshing floors and other farm buildings.

5. The word *caseiro* comes from "person who rents a *casal* (a small farm)." Because the rent is usually a share of the crop (Caldas 1981) it is appropriate to gloss caseiro as sharecropper, even though the English and Portuguese terms are not exact equivalents.

6. This emic view is in sharp contrast to that of other authors, notably Brettell (1986) and Serrão (1982) who stress emigration as a Portuguese tradition with much older roots. Some Pedralvans went to Brazil early in the century, and a few returned as Horatio Alger–type success stories, building large houses and buying farms. However, locals refer to 1964 as the beginning of emigration, meaning that emigration that year had much greater economic and social impact than earlier waves. Virtually half of Pedralva's population went to other countries in Western Europe; most earned good money and came back every year for vacation, and many returned to build big, brilliantly colored houses.

7. This is the real, inherited nickname of another, unrelated Minhoto villager and means large earthen pot. The sobriquet is unrelated to any living person's personal characteristics, at least in Pedralva.

8. This division was adopted as an attempt to be consistent with Portuguese writers. For example, Pereira (1979) divides farms into those less than one hectare, one to four hectares, four to twenty hectares, and larger categories. My own divisions could thus be aggregated to Pereira's.

9. The numbers of cattle correspond closely to amount of land per household (the correlation coefficient is .81176).

10. See Bentley 1989c for a brief biography of this man.

11. A few of the many references on Portuguese emigration include Serrão 1982; Brettell 1979, 1982, 1983, 1986; Bentley 1987b, 1989c; Goldey 1981; and Lucas 1983.

12. Stem family households are prevalent in the Mediterranean and mid-European region. Laslett (1984a) lists the Mediterranean as high-joint, low-stem family household, but the historian Rowland (1986) and his colleagues see a band of stem family households across the northern Iberian Peninsula—from at least the eighteenth century to the twentieth century—differentiating it from the neolocal latifundia areas to the south.

CHAPTER 2. AGRARIAN ECOLOGY, CULTURAL HOMOGENEITY, AND BEHAVIORAL DIVERSITY

1. Although harvesting grapes is probably as dangerous as pruning them, grape harvest comes during the peak agricultural season and many hands are needed. The vines are pruned during the winter, when there is little other farm work, making it easier to exclude women. See Pina-Cabral (1986:84) on ribald joking during grape harvest, including jests about being able to see women's legs—because they are on ladders.

2. The application of the analogy "culture is to behavior as language is to speech" must be qualified here. The natural and economic environments limit farming behavior as nothing confines speech. The limits of speech really are infinite: Pedralvans, both poor and well-off, talk about everything from the atrocities of the African wars to the marvels of Parisian shopping centers, and potentially can discuss any other topic, no matter how far removed in time, space, or imagination. However, the natural environment does not permit them to raise mangos and pineapples, just as the economic environment prevents the poorest from buying tractors and excludes all from setting up a hundred-hectare strawberry farm. Their farming behavior is infinitely varied within narrow parameters, just as there are an infinite number of fractions between zero and one.

3. Probably wireworm, the larval stage of a complex of Coleoptera species of the families Elateridae and Tenebrionidae (Ronald D. Cave, personal communication).

4. See Altieri 1984, 1986, Litsinger, Price, and Herrera 1978, and Page and Richards 1977 for descriptions of other pest-control strategies designed by peasant farmers.

5. Ellen (1982:100) argues that stone should be included with food and energy in an ecological model.

6. Coberto literally means covering. Oliveira, Galhano, and Pereira (1983: Plates 45 and 46) call such barns varandãos.

7. Figueiredo (1978) spells the word for broom codesso and identifies it as Cytysus hirsutus. Apparently codeço is an older spelling (see Michaelis 1955).

8. As Pina-Cabral (1986:87) says, the use of the term patroa is ironical, admitting and denying the power of women.

9. For descriptions of somewhat similar irrigation systems in Portugal see O'Neill 1987a:159–71, 1987b, and Pinto 1983.

10. Water may be given from other sources besides poças.

11. During field work I observed no disputes over poça water, perhaps a further suggestion that ponds are fading out of tradition, being replaced by wells. This generalization may not hold for all of the Minho, where slightly more or less water, larger or smaller pond organizations, a different water table, and any one of several other key geographic variables could change the nature of irrigation.

12. Known as milharada in some regions.

13. Restiva (Oliveira, Galhano and Perreira 1983:22); restivo (Lourenço and Alves 1968:170; Figueiredo 1978).

14. For higher fields, 395–415 meters; 375–95 meters for lower ones (with a handful as low as 330 m).

15. "Inicialmente mais pesado e puxado a gado, seguidamente de modelos mais leves e geralmente puxado por duas pessoas." The maize seeder is like a miniature plow with a wooden box on top; it plants rapidly in rows. The box is filled with maize kernels, often mixed with beans and squash seed. As a horse or cow pulls the planter a system of gears picks up the grains and drops them down a tube one by one, placing them just below the surface of the soil. Although I have seen people pulling maize planters in Beira Alta, I never observed Minhotos pull one.

16. I asked several horse owners who they had lent their animals to.

17. Strictly speaking, rego (diminutive, reguinho) means furrow. Rega (diminutive, reguinha) is the act of irrigating. I translate reguinha as little furrow because it appears more natural in English, in this context.

18. Preemergent herbicides kill broad-leafed plants, including beans and squash.

19. This generalization holds for Pedralva, and probably for the Minho as a whole, but even the smallest farmers in Honduras have a strong preference for herbicides (and chemical pest control in general). This is surprising since Central American peasant farmers have much less capital than their Portuguese counterparts. The relative supply of factors explains intracommunity variation better than cross-cultural differences. Although the sexual division of labor is much more pronounced in Honduras (few women work on farms), Honduran peasant farmers treat labor much more as a commodity than do the Portuguese. Honduran campesinos hire laborers for cash, limit

the workday to seven or eight hours, and rarely offer workers meals, coffee, or other benefits; the laborers squat at the field edges to eat the bean and tortilla *burras* they bring from home. Herbicides help Honduran farmers cut high cash expenses for hand-weeding. Minhoto farmers work from sun to sun, while Hondurans begin around dawn but quit around 2:00 P.M. often to lie exhausted in their hammocks for much of the afternoon. Perhaps influenced by plentiful food and a temperate climate, Minhoto villagers have a cultural preference for working very hard and spending little money. Honduran farmers work hard, but they periodically go hungry, have parasite burdens, are invariably thin, and they work under a harsh sun. They are more interested in substituting capital for labor (Bentley 1989a, 1991; Bentley and Andrews 1991; Andrews and Bentley 1990; DeWalt and DeWalt 1984:35).

20. It may be argued that grass hay is not palha since no part of the plant is removed for food. Seed is, however, shaken out of the grass to plant next year's crop.

21. This dangerous farm chore is one of the few reserved for men, who must climb tall, rickety ladders.

22. The finder then raised the scepterlike stalk with the red ear—the husks pulled down—shouting "I'm king, I'm king [*sou rei*]!"

23. Unfortunately, they label such work "colectivo, gratuito e recíproco" (Oliveira, Galhano and Perreira 1983:51), ignoring the inequities and exploitation of the poor that are part of labor exchange in this region (O'Neill 1982, 1987a).

24. That is, the kernels are removed from the cob.

25. Their yield of twenty-five tons (fifty tons per hectare) is very high. Yields under irrigation in Idaho averaging thirty-three tons per hectare are considered "phenomenal" (Janick et al. 1981:552).

26. *Leptinotarsa decemlineata* (Coleoptera, Chrysomelidae).

27. The diffused-light potato storage technique is perhaps the best-known example of a local practice that was formally extended by scientists to other farmers ("farmer-back-to-farmer"). According to Rhoades (1987; see also Rhoades and Booth 1982), the technique was discovered among farmers in Kenya and transferred by the Centro Internacional de la Papa, in Peru, to others in many countries. Portuguese farmers seem to have independently invented the same potato storage practice, for food stocks as well as for seed.

28. Villagers say that during the long Salazar dictatorship they were forbidden to make wine of American grapes (*vinho americano*). Rural guards would go into the back rooms of taverns, taste the wine, and empty barrels of vinho americano onto the floor. Since the revolution of 1974 Pedralvans are growing more American grapes than before; the vines are robust and yield plentiful, if somewhat inferior wine. Locals have no notion of why the grapes are called American.

29. Having lunch with a farm family one day, I got up from the table to pour myself a bowl of water; we had been drinking wine and working in the sun all morning and I was quite thirsty. The farmer called, "We have wine!" and shook his head almost disgustedly as I drank my water.

30. Usually marginal fields, at the edge of the forest, with no irrigation water and therefore ill-suited for maize.

CHAPTER 3. TECHNICAL CHANGE IN AGRICULTURE

This chapter illustrates certain ideas with case studies. For more technical details see Bentley 1987b.

1. Cancian also argues that high rank inhibits innovation, especially among the upper-middle group of farmers, who are reluctant to risk losing their status. This inhibition, combined with the facilitating effect of wealth, creates an S-shaped adoption curve exhibiting "upper-middle-class conservatism." My analysis here ignores this interesting part of Cancian's argument, both because of my small sample size and because I am interested in long-term change. Cancian (1979) claims that upper-middle-class conservatism is expressed largely in the first year or two of an innovation's introduction to a community.

2. Machinery is classed with capital.

3. A single parcel often has both forest and cropland.

4. Sometimes wool but usually linen. Flax (for making linen) was formerly intercropped with maize but was completely abandoned around 1982. Men handled the early stages of its processing. After doing the supper dishes the women and girls sat until bedtime, spinning flax thread on simple spindles and whorls, splitting their lips open by constantly moistening the thread. The women who wove the flax into linen carried the bolts of finished cloth on their heads to Guimarães, over twenty kilometers away. To avoid paying for lodgings they walked all night to reach the buyers by early morning, then hiked back home before dark. Television viewing has now replaced spinning as the late evening activity.

5. Much of this material on the period before 1964 is from informants' descriptions. Their memories generally do not go back beyond the 1940s and never beyond World War I. The depictions are subject to the distortions common with memory; people tend selectively to forget some events and embellish others. Nevertheless, these memories are worth recording because they document some information on the conditions and attitudes of rural, early-to-midtwentieth century Portuguese; they color contemporary Minhotos' attitudes about the present; and the villagers were anxious to tell me these things, and farmers and nonfarmers tended to agree on essential details such as wage rates and labor supply. (See Laslett 1984b for an excellent description of very personalized labor relations in seventeenth-century England).

6. This farm reappears in Chapter 4 as the large one-piece farm of Francisco Silva.

7. This institution reveals the etymology of the Portuguese (and Spanish) word for servant, *criado*, which also means raised. Older villagers who suffered childhood servitude say they were *criado para servir* (raised to serve).

8. As suggested by Table 3.4, the wage differences, when adjusted for inflation, are not as striking as my informants implied. No doubt villagers'

perceptions are influenced by the dramatic increase in nominal wages. Moreira (1981:298) suggests that Portuguese farmers are more influenced by the "monetary illusion" of changes in nominal prices than by changes in real prices. Another possibility is that farmers exaggerated wages more in the past, hoping that overreporting costs would lower their taxes. Government statisticians told me in 1984 that tax evaluators in Portugal used to call themselves "statisticians" as cover when talking to farmers, so that farmers generally underreported yields and exaggerated costs when reporting to survey takers. I later learned that our survey data in 1983 in Pedralva (Finan 1987) was wildly skewed. Farmers overreported cash expenses for labor by about 100 percent and underreported yields by up to 90 percent. Yet the stories of starvation and prostitution for food are too widespread to be easily dismissed (Guerreiro 1981) and support memories that wages really were quite low, and food expensive, especially during the Spanish Civil War and World War II.

9. Women are not hired as construction workers, either in the villages or the cities. On some occasions females help build their own houses, especially in the cases of young families who have not emigrated. Typically the husband is a construction worker who builds a modest home with his friends' help on days off.

10. Source is my own census of the community.

11. About U.S. $7.50. The exchange rate for the escudo was generally about ninety to the U.S. dollar in 1983 and between 120 and 130 to the dollar in 1984.

12. Women are not expressly forbidden to drive tractors, and one young woman earns her living mostly driving a tractor for the doctor farmer. But as a general rule women drive tractors only when no man is available, which is hardly ever. One woman even sold her tractor just after her husband died, saying it was because she had no man. Since about three-fourths of farm work is done by females it is very common to see one man on a tractor surrounded by a crew of women working on the ground.

13. The same household discussed earlier in this chapter, which is caring for the elderly ex-sharecropper.

14. The leaders of the cooperative believe that one advantage of this is that the money goes directly to the women of the households.

15. Renting a tractor has a high transaction cost, in economics jargon.

16. Agricultural chemicals are sold in Braga at the cooperative and in agricultural supply stores. Some taberna owners buy chemical fertilizer and repackage it for customers who want a couple of kilos for their garden.

17. Cutting brush is one of the world's most miserable jobs. The gorse is spiny and has tough roots. Great piles are cut with a hoe and mounded up—invariably driving some of the spines into one's shins. These heavy, awkward mounds are thrown onto a wagon with a pitchfork, sometimes to a height two or three meters from the ground. It is impossible to keep bits of gorse from falling down one's shirt collar. The load has to be tied on and then unloaded at the farmstead. It is no wonder that it is harder to find workers for brush cutting than for other tasks.

18. Three other reasons commonly heard for arson in northern Portugal's forest land are: (1) lucrative sales of burned and semiburned wood to timber companies; (2) destruction of wild boars (*javalís*) that prey on crops; and (3) "revenge" of villagers on the State Forestry Service for having (often decades earlier) planted forests on the poor's precious common land (*baldios*) (Brian O'Neill, personal communication).

19. Of course, we are dealing with a long period of time. It is possible that some farmers tried hybrid seed but died, moved away, or retired before I could count them. Even so, the rate of adoption was low.

20. The native land races of maize are varieties (i.e., not hybrids), and seed is selected from them every year. Grain taken from a hybrid plant (F1 generation) is viable, and some locals plant it, but its homozygosity increases every generation, and individual plants lose some desired traits.

21. This was the work of PROCALFER, a large U.S. Agency for International Development project that set up mills farther south of grind (basic) native limestone into powder for the granite-based acidic soils of the North, and that, incidentally, funded this study.

22. That is, farm size is not markedly bimodal.

23. Several government agencies and the local milk cooperatives sponsor extension agents, all of whom have their offices in the same building in Braga.

24. A few of the many citations of the failings of scientific research centers to generate relevant technology for small farmers include Bentley 1987c; Biggs 1986; Chambers 1983; Chambers and Ghildyal 1985; Chambers and Jiggins 1987; Horton 1984; Starkey 1988; Rhoades and Booth 1982.

CHAPTER 4. LAND FRAGMENTATION

1. The related phenomenon of farm fragmentation—having very small farms—is sometimes confused with land fragmentation. Brettell (1986:108) refers to nineteenth-century Portuguese households with very small amounts of land as examples of "excessive fragmentation."

2. Brettell (1986:30) remarks that in medieval Portugal much of the land was owned by the church and the aristocracy, "which contributed greatly to land fragmentation, since the more small plots that could be leased out, the greater the amount of rental income on the land an owner could collect." This enigmatic analysis seems based on a misunderstanding of the principle of fixed rent. Brettell goes on to say that "peasants, as tenant farmers, were required to pay a fixed amount of rent, usually in agricultural goods, each year, the amount of which was unrelated to the quantity of goods produced in any one year." The rents were undoubtedly related to the size of the farm and to the amount it was expected to produce in a normal, or average, year.

3. There are also a few hectares of natural pasture (known locally as *paul*) scattered about in the forest, which I have included with the monte.

4. The Swabians also practiced partible inheritance (Brettell 1986:15).

5. See also Pina-Cabral 1986:67–68. See O'Neill 1983, 1987a: ch. 7, for a discussion of this practice in nearby Trás-os-Montes. Behar (1986:98–101) reports that in León, Spain, the heir who takes on the responsibility of caring

for the elderly parents is usually rewarded with the house, in spite of the practice of partible inheritance. According to Brettell (1986:43) the terço dates to the eighteenth and nineteenth centuries in the Minho: "even within this regime of supposedly partible inheritance, one offspring was generally favored over all the others to receive the têrço [sic] or third portion."

6. Occasionally a celibate offspring remains in the house, cares for the elderly parents, and inherits the third, especially if no married children remain at home.

7. Secondary heirs, those who do not receive the third.

8. In the early 1970s one small farm of five fields was divided among six heirs. The largest field was split so each child received one small field (about 0.5 hectares each). Unable to live as farmers, all six heirs emigrated to France.

9. All names are pseudonyms.

10. The youngest generation is not depicted in the figures.

11. This is the same farm discussed in Chapter 1 where the farm family harvested grapes and ate lunch with the sharecroppers of their other farm.

12. For a description of other cases of emigration see Bentley 1989a.

13. This sample includes all households in the parish, plus the doctor's farm, minus ninety-five landless households, minus nine households with insufficient data. The land data were collected from the local "Repartição das Finanças," a manuscript copy of a 1983 land survey done by agronomists and local villagers. Additional information came from my 1984 household census (which included a land acquisition history) and from interviews with parish residents. The "Repartição das Finanças" land survey listed each parcel by a named geographical area, indicating size and other information. The household census elicited number of parcels rented and bought by date of acquisition. Follow-up interviews with informants confirmed the location of parcels that had been bought and rented, allowing me to calculate parcel and farm size.

14. Sharecroppers are lumped into the rental category because there are only six of them, and villagers perceive their cultural importance and numbers dwindling (Caldas 1981). Landowners wishing to retain sharecroppers now must offer inducements such as permitting them to keep small dairy herds and enjoy the dairy profits. In 1984 Pedralva's sharecroppers farmed seventeen fields (average of 2.8 fields per household), totaling 7.138 hectares (average of 1.1897 ha of cropland per household and 0.4199 ha per field). The number and size of these plots suggest that sharecropper farms are less fragmented than other farms of similar size (see Table 4.2). This is a counterexample to Brettell's (1986:108) notion that rental contracts in the nineteenth century were "passed from one generation to the next, thereby creating further subdivisions."

15. The square root of the total farm area, divided by the sum of the square roots of the plot sizes. A one-piece farm has a value of 1, and more-fragmented farms have values tending toward zero. A property of this index is that fragmentation decreases (the index value approaches 1) when the area of big plots increases and that of small plots decreases. Januszewski's index

measures number of plots and size distribution of plots (King and Burton 1982:476).

16. The sum of the one-way road distance from the farmstead to each field; its value increases as the number of fields increases and as the distance to them increases. This measure gives a rough idea of the distance involved on a fragmented farm, but it ignores field size and groupings of fields. A line drawn on a map around most farms with distant fields would be elliptical rather than circular, because most farms with distant fields have their distant fields clustered in one small area.

17. The area (in ha) of each field multiplied by the one-way road distance (in km) from the farmstead; the sum of the value for each plot is the value for the individual farm. This is a modification of a concept developed by Igbozurike (1974). Intuitively, a farm with a one-hectare field two kilometers from the farmstead and a thousand-square-meter plot next to the house is more fragmented than a farm with a thousand-square-meter plot at two kilometers and a one-hectare plot out the back door, because the farmer whose larger plot is more distant must make more frequent trips and must haul more supplies and harvests over the two kilometers than the farmer with the smaller field at two kilometers. This index increases as the larger fields are farther away. Another advantage is that a group of fields in a single location would have the same effect on the total score as a single field (of the same total size) at the same location. A farmer with several fields in a single location can visit each without returning to the farmstead.

18. A 1655 essay by Manuel Severim de Faria, cited by Brettell (1986:75) claims that "the bigger the estate, the more it is divided into small parcels."

19. Carvalho, Barros, and Rocha (1982) report that in Beira Litoral in 1970 milk yields per cow on large farms were higher than those on small farms by 789 kilograms in the first lactation, 1,108 in the third, and 1,080 in the sixth. But by 1978 the small farmers had caught up with the large farmers technologically and were outproducing them by 191, 538, and 244 kilograms, respectively (p. 142). Smaller farmers also have lower cash expenses per liter of milk (p. 150).

20. Critics might correctly observe that an insignificant F-value does not prove a lack of correlation, since the F-value is modeled after a straight line. A yield increase for large farmers or for consolidated farms on a steep curve could also produce a low F-value, although the correlation would be significant. Scattergram analysis of the same data indicated absolutely no correlation between yield and fragmentation.

CHAPTER 5. LAND USE CHANGES AND THE ECOLOGY
OF FIELD AND FOREST

1. The large farmhouses commonly have names, often a place-name (e.g., Vila Nova) or an ancestor's name. People are much more widely known by their house name than by their legal name. For example, the people who live in Vila Nova might be called José (or Maria) de Vila Nova, except for the (male) household head, who would simply be called Vila Nova.

2. The same attitude is prevalent in Trás-os-Montes toward plots left uncultivated (*deixados à monte*). Smallholders and middle peasants, when talking of large landowners, stress the unnecessary "waste" of theoretically cultivable land (O'Neill, personal communication).

CHAPTER 6. THE EARTH HAS BONES

1. Based on how closely villagers count anything that is worth money, I suspect that this is somewhat a polite fiction. People probably do keep track of their exchanges, privately, but as long as they are not too far out of balance no one complains.

CHAPTER 7. THE FUTURE OF SMALL-SCALE FARMING IN NORTHEAST PORTUGAL

1. O'Neill (personal communication) claims that the agriculture of Trás-os-Montes (in the Northeast) is much less sophisticated than that of the Minho (in the Northwest), and that "poor technology" refers to the "predominance of ox carts, sickles, scythes, hoes, picks, and ubiquitous unpaid work groups." Nevertheless, "poor technology" is an unfortunate choice of words to describe an agricultural system that has survived for a long time and is therefore by definition sustainable, and that also requires few outside inputs (especially chemicals) and is therefore probably agroecologically sound (see Altieri 1987).

References

ACEVES, JOSEPH B.
 1971 *Social change in a Spanish village.* Cambridge, Mass: Schenkman.

ALTIERI, MIGUEL A.
 1984 Desarrollo de estrategias para el manejo de plagas por campesinos, basándose en el conocimiento tradicional. *CIRPON—revista de investigación* 2 (3–4):151–64.
 1986 Bases ecológicas para el desarrollo de sistemas agrícolas alternativos para campesinos de Latinoamérica. *CIRPON* 4 (1–4):83–109.
 1987 *Agroecology: The scientific basis of alternative agriculture.* Boulder, Colo.: Westview.

ANDREWS, KEITH L., AND JEFFERY W. BENTLEY
 1990 IPM and resource-poor Central American farmers. *Global Pesticide Monitor* 1 (2 [May]):1, 7–9.

BARLETT, PEGGY F.
 1982 *Agricultural choice and change: Decision making in a Costa Rican community.* New Brunswick, N.J.: Rutgers University Press.

BEHAR, RUTH
 1986 *Santa María del Monte: The presence of the past in a Spanish village.* Princeton, N.J.: Princeton University Press.

BENTLEY, JEFFERY W.
 1987a Economic and ecological approaches to land fragmentation. *Annual Review of Anthropology* 16:31–67.
 1987b Technical change in a northwest parish. In *Portuguese agriculture in transition,* ed. Scott R. Pearson et al. Ithaca, N.Y.: Cornell University Press.
 1987c Water harvesting on the Papago Reservation: Experimental agricultural technology in the guise of development. *Human Organization* 46 (2):141–46.

1989a   Pérdida de confianza en conocimiento tradicional como resultado de extensión agrícola entre campesinos del sector reformado en Honduras. *Ceiba* 30 (1).

1989b   What farmers don't know can't help them: The strengths and weaknesses of indigenous technical knowledge in Honduras. *Agriculture and Human Values* 6 (3 [Summer]):25–31.

1989c   Eating the dead chicken: Intra-household decision making and emigration in rural Portugal. In *The household economy: Reconsidering the domestic mode of production*, ed. Richard R. Wilk. Boulder, Colo.: Westview.

1990   Conocimiento y experimentos espontáneos de campesinos hondureños sobre el maíz muerto. *Manejo Integrado de Plagas* 17:16–26.

1991   ¿Qué es hielo? percepciones de los campesinos hondureños sobre enfermedades del frijol y otros cultivos. *Interciencia* 16(3):131–37.

BENTLEY, JEFFERY W., AND KEITH L. ANDREWS

1991   Pests, peasants and publications: Anthropological and entomological views of an integrated pest management program for small-scale Honduran farmers. *Human Organization* 50(2):113–24

BIGGS, STEPHEN D.

1986   Agricultural technology generation and diffusion: Lessons for research policy. Agricultural Administration (Research and Extension) Network Discussion Paper, no. 16. London: Overseas Development Institute.

1989   A multiple source of innovation model of agricultural research and technology promotion. Agricultural Administration (Research and Extension) Network Paper, no. 6. London: Overseas Development Institute.

BINNS, B. O.

1950   *The consolidation of fragmented agricultural holdings.* FAO Agricultural Studies, no. 11. Washington, D.C.: Food and Agriculture Organization of the United Nations.

BIRCH, THOMAS W.

1986   Communicating with nonindustrial private forestland owners. *Journal of Forestry* 84 (12):25–33.

BLAIKIE, P.

1971   Spatial organization of agriculture in some north Indian villages: Part 1. *Transactions of the Institute of British Geographers* 52:1–40.

BLOCH, MARC

1966   *French rural history: An essay on its basic characteristics.* Berkeley: University of California Press.

BLOK, ANTON

1981   Carneiros e cabrões—uma oposição chave para o código mediterrâneo de honra. *Perspectives sobre o Norte de Portugal. Estudos contemporâneos* 2/3:9–30.

BOURDIEU, PIERRE
1976   Marriage strategies as strategies of social reproduction. In *Family and society*, ed. R. Forster and O. Ranum. Baltimore, Md.: Johns Hopkins University Press.

BOX, LOUK
1988   Experimenting cultivators: A method for adaptive agricultural research. *Sociologia Ruralis* 28:62–75.

BRAMMER, HUGH
1980   Some innovations don't wait for experts: A report on applied research by Bangladeshi peasants. *Ceres* 13 (2):24–28.

BRETTELL, CAROLINE B.
1979   Emigrar para voltar: A Portuguese ideology of return migration. *Papers in Anthropology* 20(1): 1–20.
1982   *We have already cried many tears: The stories of three Portuguese migrant women*. Boston: Schenkman.
1983   Emigração, a igreja e a festa religiosa do Norte de Portugal: estudo de um caso. *Comunidades rurais: Estudos interdisciplinares. Estudos contemporâneos* 5:175–204.
1986   *Men who migrate, women who wait: Population and history in a Portuguese parish*. Princeton, N.J.: Princeton University Press.

BROWN, PAULA, AND AARON PODOLEFSKY
1976   Population density, agricultural intensity, land tenure and group size in the New Guinea highlands. *Ethnology* 15:211–38.

BURTON, STEVE, AND RUSSEL KING
1982   Land fragmentation and consolidation in Cyprus. *Agricultural Administration* 11 (3):183–200.

CALDAS, JOÃO CASTRO
1981   Caseiros de Alto Minho: adaptação e declínio. *A pequena agricultura em Portugal. Revista crítica de ciências sociais* 7/8:203–16.

CANCIAN, FRANK
1972   *Change and uncertainty in a peasant economy: The Maya corn farmers of Zinacantán*. Stanford, Calif.: Stanford University Press.
1979   *The innovator's situation: Upper-middle-class conservatism in agricultural communities*. Stanford, Calif.: Stanford University Press.

CARLYLE, WILLIAM J.
1983   Fragmentation and consolidation in Manitoba. *The Canadian Geographer* 27 (1):17–34.

CARVALHO, AGOSTINHO DE, VÍTOR COELHO BARROS, AND JOSÉ RAMOS ROCHA
1982   *Que futuro para a produção leiteira: grande ou pequena exploração*. Oeiras, Portugal: Instituto Gulbenkian de Ciência, Centro de Estudos de Economia Agrária.

CHAMBERS, ROBERT
1983   *Rural development: Putting the last first*. New York: Wiley and Sons.

CHAMBERS, ROBERT, AND B. P. GHILDYAL
1985   Agricultural research for resource-poor farmers: The farmer-first-and-last model. *Agricultural Administration* 20:1–30.

CHAMBERS, ROBERT, AND JANICE JIGGINS
1987a   Agricultural research for resource-poor farmers. Part 1: Transfer-of-technology and farming systems research. *Agricultural Administration and Extension* 27:35–52.
1987b   Agricultural research for resource-poor farmers. Part 2: A parsimonious paradigm. *Agricultural Administration and Extension* 27:109–28.

CHISHOLM, MICHAEL
1979   *Rural settlement and land use: An essay in location.* London: Hutchinson.

CHRISTIAN, WILLIAM
1972   *Person and God in a Spanish valley.* New York: Seminar.

CLOUT, H. D.
1972a   *Agriculture.* London: Macmillan Press.
1972b   *Rural Geography.* Oxford: Pergamon.

COLE, J. W., AND E. R. WOLF
1974   *The hidden frontier: Ecology and ethnicity in an Alpine valley.* New York: Academic Press.

CUTILEIRO, JOSÉ
1971   *A Portuguese rural society.* Oxford: Clarendon Press.

DAVIS, JOHN H. R.
1977   *People of the Mediterranean: An essay in comparative social anthropology.* London: Routledge and Kegan Paul.

DELISLE, DAVID
1982   Effects of distance on cropping patterns internal to the farm. *Annals of the Association of American Geographers* 72 (1):88–98.

DEWALT, BILLIE R., AND KATHLEEN M. DEWALT
1984   *Sistemas de cultivos en Pespire, sur de Honduras: un enfoque de agroecosistemas.* Tegucigalpa: Instituto Hondureño de Antropología e Historia.

DIAS, JORGE
1981   *Rio de Onor: comunitarismo agro-pastoril.* 2d ed. Lisbon: Editorial Presença, orig. pub. 1953.

DOUGLASS, WILLIAM A.
1969   *Death in Murélaga: Funerary ritual in a Spanish Basque village.* Seattle: University of Washington Press.
1975   *Echalar and Murelaga: Opportunity and rural exodus in two Spanish Basque villages.* New York: St. Martin's Press.

DOVRING, FOLKE
1965   *Land and labour in Europe in the twentieth century.* 3d ed. The Hague: Nijhoff.

DOWNING, THEODORE
1977   Partible inheritance and land fragmentation in a Oaxaca village. *Human Organization* 36:235–43.

THE ECONOMIST
1984   Time to cooperate: Agriculture needs all the time and help it can get from the ECC. *The Economist* 291 (7348 [June 30]).

EDWARDS, C.J.W.
1978   The effects of changing farm size upon levels of farm fragmentation: A Somerset case study. *Journal of Agricultural Economics* 29 (2):143–54.

ELLEN, ROY F.
1982   *Environment, subsistence and system: The ecology of small-scale social formations.* New York: Cambridge University Press.

FARMER, B. H.
1960   On not controlling subdivision in paddy lands. *Transactions of the Institute of British Geographers* 28:225–35.

FENOALTEA, STEFANO
1976   Risk, transaction costs and the origin of medieval agriculture. *Explanations in Economic History* 13:129–51.

FIGUEIREDO, CÂNDIDO DE
1978   *Pequeno dicionário da língua portuguesa.* Lisbon: Livraria Bertrand.

FINAN, TIMOTHY J.
1987   Intensive agriculture in the Northwest. In *Portuguese agriculture in transition,* ed. Scott R. Pearson et al. Ithaca, N.Y.: Cornell University Press.

FORBES, H. A.
1976   'We have a little of everything': The ecological basis of some agricultural practices in Methana, Trizinia. In *Regional variation in modern Greece and Cyprus: Towards a perspective on the ethnoography of Greece,* ed. M. Dimen and E. Friedl, a special edition of *Annals of the New York Academy of Sciences,* no. 268.

FOX, ROGER, AND TIMOTHY J. FINAN
1987a   Patterns of technical change in the Northwest. In *Portuguese agriculture in transition,* ed. Scott R. Pearson et al. Ithaca, N.Y.: Cornell University Press.
1987b   Future technical and structural adjustments in northwestern agriculture. In *Portuguese agriculture in transition,* ed. Scott R. Pearson et al. Ithaca, N.Y.: Cornell University Press.

FRIEDL, J.
1974   *Kippel: A changing village in the Alps.* New York: Holt, Rinehart, and Winston.

GALT, A. H.
1979   Exploring the cultural ecology of field fragmentation and scattering in the Island of Pantellaria. *Journal of Anthropological Research* 35:93–108.

GEERTZ, CLIFFORD
1963   *Agricultural involution.* Berkeley: University of California Press.

GOFFMAN, ERVING

1959   *The presentation of self in everyday life.* New York: Doubleday Anchor.

1967   *Interaction ritual.* Chicago: Aldine.

GOLDEY, PATRICIA

1981   Emigração e estrutura familiar—estudo de um caso no Minho. *Perspectivas sobre o Norte de Portugal. Estudos contemporâneos* 2/3:111–27.

GOODENOUGH, WARD

1971   *Culture, language and society.* McCaleb Module in Anthropology. Reading, Mass.: Addison-Wesley.

GRIGG, DAVID

1980   *Population growth and agrarian change: An historical perspective.* Cambridge: Cambridge University Press.

1983   Agricultural geography. *Progress in Human Geography* 7 (2):255–60.

GUERREIRO, MANUEL VIEGAS

1981   *Pitões das Júnias: esboço de monografia etnográfica.* Lisbon: Serviço Nacional de Parques, Reservas e Património Paisagístico.

GUICHARD, FRANÇOIS

1982   *Atlas demográfico de Portugal.* Lisbon: Livros Horizonte.

GUILLET, DAVID

1981   Land tenure, ecological zone, and agricultural regime in the central Andes. *American Ethnologist* 8 (1):139–56.

HANKS, LUCIEN

1972   *Rice and man: Agricultural ecology in Southwest Asia.* Chicago: Aldine.

HARRIS, MARVIN

1985   *Culture, people, nature.* 4th ed. New York: Harper and Row.

HERZFELD, M.

1980   Social tension and inheritance by lot in three Greek villages. *Anthropological Quarterly* 53:91–100.

HESTON, ALAN, AND DHARMA KUMAR

1983   The persistence of land fragmentation in peasant agriculture: An analysis of south Asian Cases. *Explanations in Economic History* 20 (2):199–220.

HILLIARD-CLARK, JOYCE, AND CLYDE E. CHESNEY

1985   Black woodland owners: A profile. *Journal of Forestry* 83 (11): 674–79.

HOLMES, DOUGLAS R.

1983   A peasant-worker model in a northern Italian context. *American Ethnologist* 10 (4):734–48.

HORTON, DOUGLAS E.

1984   *Social scientists in agricultural research: Lessons from the Mantaro Valley Project, Peru.* Ottawa: International Development Research Centre.

1986   Farming systems research: Twelve lessons from the Montaro Valley Project. *Agricultural Administration* 23:93–107.

HYODO, SETSURO
1956   Aspects of land consolidation in Japan. In *Land tenure,* ed. Kenneth H. Parsons, Raymond J. Penn, and Philip M. Raup. Madison: University of Wisconsin Press.

IGBOZURIKE, M. U.
1970   Fragmentation in tropical Africa: An overrated phenomenon. *Professional Geographer* 22:132–35.
1974   Land tenure, social relations and the analysis of spatial discontinuity. *Area* 6:132–36.

ILBERY, BRIAN W.
1984   Farm fragmentation in the Vale of Evesham. *Area* 16 (2):159–65.

ITURRA, RAÚL
1980   Strategies in the domestic organization of production in rural Galicia (N.W. Spain). *Cambridge Anthropology* (double issue: *Studies in European Ethnography*) 6 (1–2):88–129.
1983   Estratégias na organização doméstica da produção na Galiza rural. *Ler historia* 1:81–109.

JACKSON, J. P.
1984   Nonindustrial private forests: New look at an old problem. *American Forests* 90 (2):25–27, 54.

JACKSON, R. T.
1970   Some observations on the Von Thünen method of analysis with reference to southern Ethiopia. *East African Geographical Review* 8:39–46.

JACOBY, E. H.
1971   *Man and land.* London: Andre Deutsch.

JANICK, JULES ET AL.
1981   *Plant science: An introduction to world crops.* 3d ed. San Francisco: W. H. Freeman.

JOHNSON, ALLEN W.
1971   *Sharecroppers of the Sertão: Economics and dependence on a Brazilian plantation.* Stanford, Calif.: Stanford University Press.
1972   Individuality and experimentation in traditional agriculture. *Human Ecology* 1 (2):149–59.

JOHNSON, O.E.G.
1970   A note on the economics of fragmentation. *Nigerian Journal of Economic and Social Studies* 12:175–84.

KAROUZIS, G.
1971   Time wasted and distance travelled by the average Cypriot farmer in order to visit his scattered and fragmented agricultural holding. *Geographical Chronicles* 1:39–58.

KERR, WARWICK ESTEVAM, AND DARRELL ADDISON POSEY
1984   Informações adicionais sobre a agricultura dos Kayapó. *Interciencia* 9:392–400.

KING, R. L.
1977   *Land reform: A world survey.* London: G. Bell and Sons.

KING, R. L., AND S. P. BURTON
1982   Land fragmentation, a fundamental rural spatial problem. *Progress in Human Geography* 6:475–94.
1983   Structural change in agriculture: The geography of land consolidation. *Progress in Human Geography* 7 (4):471–501.

KOGAN, MARCOS
1975   Plant resistance in pest management. In *Introduction to insect pest management,* ed. Robert Metcalf and William Luckman. New York: Wiley and Sons.

LAMBERT, A. M.
1963   Farm consolidation in Western Europe. *Geography* 48:31–48.

LASLETT, PETER
1984a   The family as a knot of individual differences. In *Households: Comparative and historical studies of the domestic group,* ed. Robert M. Netting, Richard R. Wilk, and Eric J. Arnould. Berkeley: University of California Press.
1984b   *The world we have lost.* New York: Scribner.

LEACH, EDMUND
1968   *Pul Eliya: A village in Ceylon.* Cambridge: Cambridge University Press.

LIGHTFOOT, CLIVE
1987   Indigenous research and on-farm trials. *Agricultural Administration and Extension* 24:79–89.

LIPTON, MICHAEL
1964   Population, land and decreasing returns to agricultural labour. *Bulletin of the Oxford University Institute of Economics and Statistics* 26 (2):123–57.

LITSINGER, J. A., E. C. PRICE, AND R. T. HERRERA
1978   Filipino farmer use of plant parts to control rice insect pests. *International Rice Research Newsletter* 3 (5):15–16.

LOURENÇO, J. SILVA, AND VÍTOR MANUEL ALVES
1968   *Tempos de trabalho agrícola numa região do Noroeste.* Lisbon: Fundação Calouste Gulbenkian, Centro de Estudos de Economia Agrária.

LUCAS, ANTÓNIO M. ROLO
1983   O lugar da palhaça e a feira dos quatro caminhos. *Comunidades rurais: Estudos interdisciplinares. Estudos contemporâneos* 5:151–73.

LYNN SMITH, T.
1959   Fragmentation of agricultural holdings in Spain. *Rural Sociology* 24:140–49.

MCCLOSKEY, DONALD N.
1975   The persistence of English open fields. In *European peasants and their markets: Essays in agrarian economic history,* ed. William Parker and E. L. Jones. Princeton, N.J.: Princeton University Press.

1976   English open fields as behavior towards risk. In *Research in economic history*, ed. P. J. Uselding. Greenwich, Conn.: JAI Press.

MELICZEK, H.

1973   The work of FAO and experiences in land consolidation. *Land Reform, Land Settlement and Cooperatives* 1:50–64.

MICHAELIS, H.

1955   *Dictionary of the Portuguese and English languages.* New York: Frederick Ungar.

MONKE, ERIC

1987a   Agricultural factor markets. In *Portuguese agriculture in transition*, ed. Scott R. Pearson et al. Ithaca, N.Y.:Cornell University Press.

1987b   Future policies influencing agricultural factor markets. In *Portuguese agriculture in transition*, ed. Scott R. Pearson et al. Ithaca, N.Y.: Cornell University Press.

MOORE, WILBERT E.

1972   *Economic demography of Eastern and southern Europe.* New York: Arno Press.

MOREIRA, MANUEL BELO

1981   A pequena produção e os estímulos do mercado: o caso da produção do leite no Concelho de Vagos. *A pequena agricultura em Portugal*, a special issue of *Revista crítica de ciências sociais* 7/8: 289–308.

NAYLON, J.

1959   Land consolidation in Spain. *Annals of the Association of American Geographers* 49:361–73.

NETTING, ROBERT McC.

1969   Ecosystems in process: A comparative study of change in two West African societies. In *Contributions to anthropology: Ecological essays*, ed. David Damas. Bulletin, no. 230. Ottawa: National Museum of Canada.

1972   Of men and meadows: Strategies of Alpine land use. *Anthropological Quarterly* 45:132–44.

1974a   Agrarian ecology. *Annual Review of Anthropology* 3:21–56.

1974b   The system nobody knows: Village irrigation in the Swiss Alps. In *Irrigation's impact on society*, ed. T. E. Downing and M. Gibson. Tucson: University of Arizona Press.

1976   What Alpine peasants have in common: Observations on communal tenure in a Swiss village. *Human Ecology* 4:135–46.

1981   *Balancing on an Alp: Ecological change and continuity in a Swiss mountain community.* New York: Cambridge University Press.

1982a   Territory, property and tenure. In *Behavioral and social science research: A national resource*, ed. Robert McC. Adams, Neil J. Smelser, and Donald J. Treiman. Washington, D.C.: National Academy Press.

1982b   Some home truths on household size and wealth. *American Behaviorial Scientist* 25 (6):641–62.

OECD (ORGANIZATION FOR ECONOMIC COOPERATION AND DEVELOPMENT)
1964   *Low incomes in agriculture—problems and policies.* Paris: OECD Agricultural Policy Report.
1969   *Agricultural development in southern Europe.* Paris: OECD Agricultural Policy Report.
1972   *Structural reform measures in agriculture.* Paris: OECD Agricultural Policy Report.

O'FLANAGAN, T. P.
1980   Agrarian structures in northwestern Iberia: Responses and their implications for development. *Geoforum* 11:157–69.

OLIVEIRA, ERNESTO VEIGA DE, FERNANDO GALHANO, AND BENJAMIN PEREIRA
1983   *Afaia agrícola portuguesa.* 2d ed. Lisbon: Instituto de Investigação Científica.

O'NEILL, BRIAN JUAN
1981   Propietários, jornaleiros e criados numa aldeia transmontana desde 1886. *Perspectivas sobre o Norte de Portugal. Estudos contemporâneos* 2/3:31–73.
1982   Trabalho cooperativo numa aldeia do Norte de Portugal. *Análise social* 18:7–34.
1983   Dying and inheriting in rural Trás-os-Montes. *Journal of the Anthropological Society of Oxford* 14 (1):44–74.
1984   *Propietários, lavradores e jornaleiras: desigualdade social numa aldeia transmontana 1870–1978.* Lisbon: Publicações Dom Quixote.
1987a   *Social inequality in a Portuguese hamlet: Land, late marriage and bastardy, 1870–1978.* Cambridge: Cambridge University Press.
1987b   *Pul Eliya* in the Portuguese mountains: A comparative essay on kinship practices and family ideology. *Sociologia Ruralis* 27 (4): 278–303.

PAGE, W. W., AND P. RICHARDS
1977   Agricultural pest control by community action: The case of the variegated grasshopper in southern Nigeria. *African Environment* 2/3:127–41.

PEARSON, SCOTT R.
1987   Portuguese agricultural strategies. In *Portuguese agriculture in transition,* ed. Scott R. Pearson et al. Ithaca, N.Y.: Cornell University Press.

PEARSON, SCOTT R., FRANCISCO AVILLEZ, JEFFERY W. BENTLEY, TIMOTHY J. FINAN, TIMOTHY JOSLING, MARK LANGWORTHY, ERIC MONKE, AND STEFAN TANGERMANN
1987   *Portuguese agriculture in transition.* Ithaca, N.Y.: Cornell University Press.

PEARSON, SCOTT R., AND ERIC MONKE
1987   Constraints on the development of Portuguese agriculture. In *Portuguese agriculture in transition,* ed. Scott R. Pearson et al. Ithaca, N.Y.: Cornell University Press.

PEREIRA, MARIO
1979 *A estrutura agrária portuguesa (1968–1970): suas Relações* com a população e a produção agrícolas. Oeiras: Instituto Gulbenkian de Ciência.

PHLIPPONNEAU, MICHEL
1975 Breton farmyard politics. *Geographical Magazine* 47 (5):289–95.

PINA-CABRAL, JOÃO DE
1986 *Sons of Adam, daughters of Eve: The peasant worldview of the Alto Minho.* Oxford: Clarendon Press.

PINTO, MANUEL
1983 Da agua de rega à agua ritual (apontamentos sobre o caso da freguesia de Sobrado-Valongo). *Comunidades rurais, Estudos interdisciplinares. Estudos contemporâneos* 5:117–49.

PITT-RIVERS, JULIAN A.
1961 *The people of the Sierra.* Chicago: University of Chicago Press.

PORTELA, JOSÉ
1981 Fragueiro: notas sobre a agricultura local. A Pequena Agricultura em Portugal. *Revista crítica de ciências sociais* 7/8:217–46.

PORTUGAL. INSTITUTO NACIONAL DE ESTATÍSTICA
1930, *Recenseamento geral da população.* Lisbon: Instituto Nacional
1940, de Estatística.
1960,
1980

REDFIELD, ROBERT E.
1960 *Peasant society and culture.* Chicago: University of Chicago Press.

RHOADES, ROBERT E.
1979 From caves to Main Street: Return migration and the transformation of a Spanish village. *Papers in Anthropology* 20 (1):57–74.
1980 European cyclical migration and economic development: The case of Spain. In *Urban life: Studies in urban anthropology,* ed. George Gmelch and Walter P. Zenner. New York: St. Martin's Press.
1987 Farmers and experimentation. Agricultural Administration Unit Discussion Paper, no. 21. London: Overseas Development Institute.

RHOADES, ROBERT E., AND ANTHONY BEBBINGTON
1988 Farmers who experiment: An untapped resource for agricultural research and development. Paper presented at the International Congress of Plant Physiology, New Delhi, India, February 15–20.

RHOADES, ROBERT E., AND ROBERT H. BOOTH
1982 Farmer-back-to-farmer: A model for generating acceptable agricultural technology. *Agricultural Administration* 11:127–37.

RHOADES, ROBERT E., AND S. I. THOMPSON
1975 Adaptive strategies in Alpine environments: Beyond ecological particularism. *American Ethnologist* 2:535–51.

RICHARDS, PAUL

1985   *Indigenous agricultural revolution: Ecology and food production in West Africa.* Boulder, Colo.: Westview Press.

1986   *Coping with hunger: Hazard and experimentation in an African rice-farming system.* London: Allen and Unwin.

1989a   Agriculture as a performance. In *Farmer first: Farmer innovation and agricultural research,* ed. Robert Chambers, Arnold Pacey, and Lori Ann Thrupp. London: Intermediate Technology Publications.

1989b   Farmers also experiment: A neglected intellectual resource in African science. *Discovery and Innovation* 1 (1):19–25.

ROMM, JEFF, RAUL TUAZON, AND COURTLAND WASHBURN

1987   Relating forestry investment to characteristics of nonindustrial private forestland owners in northern California. *Forest Science* 33 (1):197–209.

ROSCH, ELEANOR

1978   Principles of categorization. In *Cognition and categorization,* ed. Eleanor Rosch and Barbara B. Lloyd. Hillsdale, N.J.: Lawrence Erlbaum Associates.

ROWLAND, ROBERT

1986   Sistemas matrimoniales en la Península Ibérica (siglos XVI–XIX): una perspectiva regional. In *La demografía histórica de la Península Ibérica,* ed. V. Pérez Moreda and D. S. Reher. Actas de las I Jornadas de Demografía Histórica, Madrid, December 1983. Madrid: Editorial Tecnos.

RUTTAN, VERNON W., AND YUJIRO HAYAMI

1984   Induced innovation model of agricultural development. In *Agricultural development in the Third World,* ed. Carl K. Eicher and John M. Staatz. Baltimore, Md.: Johns Hopkins University Press.

SCHMOOK, G., JR.

1976   The spontaneous evolution from farming on scattered strips to farming in severality in Flanders between the sixteenth and twentieth centuries: A quantitative approach to the study of farm fragmentation. In *Fields, farms and settlement in Europe,* ed. R. H. Buchanan, R. A. Butlin, and D. McCourt. Belfast: Ulster Folk and Transport Museum.

SERRÃO, JOEL

1982   *A emigração portuguesa.* 4th ed. Lisbon: Livros Horizonte.

SILVA, ROSA FERNANDA MOREIRA DA

1983   Contraste e mutações na paisagem agrária das planícies e colinas minhotas. *Comunidades rurais, Estudos interdisciplinares. Estudos contemporâneos* 5:9–115.

SMITH, C. T.

1978   *An historical geography of Western Europe before 1800.* New York: Longman Press.

STANISLAWSKI, DAN

1959   *The individuality of Portugal: A study in historical-political geography.* Austin: University of Texas Press.

1970   *Landscapes of Bacchus: The vine in Portugal.* Austin: University of Texas Press.

STARKEY, PAUL

1988   Practical agricultural research: Lessons from thirty years of developing wheeled toolcarriers. Agricultural Administration (Research and Extension) Network Discussion Paper, no. 25. London: Overseas Development Institute.

THOMPSON, K.

1963   *Farm fragmentation in Greece.* Research Monograph Series, no. 5. Athens: Centre for Economic Research.

TOWELL, WILLIAM E.

1982   Managing private nonindustrial forestlands: A perennial issue. *Journal of Forest History* 26:192–97.

UDO, RUBEN

1965   Disintegration of nucleated settlement in eastern Nigeria. *Geographical Review* 55:53–67.

VANDER MEER, P.

1975   Land consolidation through land fragmentation: Case studies from Taiwan. *Land Economics* 51:275–83.

VON DIETZ, CONSTANTIN C.

1956   Land consolidation procedures: A comparative analysis. In *Land tenure*, ed. Kenneth H. Parsons, Raymond J. Penn, and Philip M. Raup. Madison: University of Wisconsin Press.

WEINBERG, D.

1972   Cutting the pie in the Swiss Alps. *Anthropological Quarterly* 45:125–31.

WILLEMS, EMILIO

1962   On Portguese family structure. *International Journal of Comparative Sociology* 3:65–79.

YOUNG, ROBERT A., AND MICHAEL R. REICHENBACH

1987   Factors influencing the timber harvest intentions of nonindustrial private forest owners. *Forest Science* 33 (2):381–92.

ZABAWA, ROBERT

1987   Macro-micro linkages and structural transformation: The move from full-time to part-time farming in a north Florida agricultural community. *American Anthropologist* 89 (2):366–82.

Index

ABOUT THE AUTHOR

Jeffery W. Bentley has worked as an anthropologist for the Crop Protection Department at the Escuela Agrícola Panamericana, El Zamorano, Honduras, since 1987. He received his Ph.D. in cultural anthropology from the University of Arizona in 1986. He has published numerous articles on topics including land fragmentation, technical change in agriculture, water harvesting, campesinos' perceptions of plant diseases, and integrated pest management (IPM). He is currently working on the ethnoentomology of wasps and ants, with a focus on the possibility of using them as an alternative to chemical insecticides in Central America.